George Smith Heatley

Practical veterinary remedies: A useful handbook on medicine

Describing its properties, action, uses and doses

George Smith Heatley

Practical veterinary remedies: A useful handbook on medicine
Describing its properties, action, uses and doses

ISBN/EAN: 9783337197384

Printed in Europe, USA, Canada, Australia, Japan

Cover: Foto ©berggeist007 / pixelio.de

More available books at **www.hansebooks.com**

PRACTICAL

VETERINARY REMEDIES

A

USEFUL HAND-BOOK ON MEDICINE.

Describing its Properties, Action, Uses, and Doses, together with
Instructions How to Administer it to the Horse,
Ox, Cow, Sheep, Pig, and Dog.

BY

G. S. HEATLEY, M.R.C.V.S.,

AUTHOR OF "THE HORSE OWNER'S SAFEGUARD," "THE STOCK
OWNER'S GUIDE," "SHEEP FARMING," "OUR DOGS
AND THEIR DISEASES," ETC. ETC.

EDINBURGH:

MACLACHLAN & STEWART,

(Booksellers to the University.)

LONDON: BAILLIERE, TINDALL, & CO.
NEW YORK: WILLIAM R. JENKINS.

1885.

PREFACE.

THE ostensible object of the present treatise is to present to all those interested in the welfare and management of the domestic animals herein mentioned a reliable guide concerning "Practical Veterinary Remedies."

It is therefore anticipated that the work will be found useful and important to the numerous constituency to which it appeals, and that the truisms it contains may be the means of mitigating much suffering, besides tending to keep down inordinate expense.

The Author also wishes to express his heartfelt gratitude to the Press for the very favourable consideration he has already received for former efforts in this class of literature.

March 1885.

PRACTICAL VETERINARY REMEDIES.

INTRODUCTORY.

ENCOURAGED by the success attending my former efforts in the scientific literary arena, and yielding to the earnest entreaties of numerous friends throughout the country, I have at length produced, with the assistance of the publishers, a cheap book, containing a number of pertinent facts relating to Practical Veterinary Remedies.

In doing so, we wish not to usurp the place of others who have faithfully chronicled their views ; but as many a valuable sermon can be preached from the same text, yet clothed in more attractive language, so many a work may be written even upon the same foundation. Not that we entertain egotistical ideas regarding our own productions, but because we conscientiously aim at placing before the reader a few truisms that are of everyday importance in the administration of medicines to the lower animals.

How often have we witnessed the career of a valuable animal abruptly terminated by an injudicious dose of medicine !

I well remember being requested to visit a horse said to be suffering from colic. On my arrival I found the symptoms of colic accurate enough, with the complication of congestion of the lungs. Now I was rather at a loss to account for both maladies being present at one and the same time, so I proceeded to interrogate the attendant as follows :—

"What has this horse been doing, John ?"

"Oh, naething but plooing, sir."

"Has he been feeding all right ?"

"Yes, up to this morning."

"Did you observe him blowing or lifting at the flanks yesterday ?"

"No ; he was a' richt last nicht when we left the stable, but this morning he wud neither eat nor drink ; and as he looked as if he was griped, we gave him a bottle o' linseed oil, and a tablespoonful o' ginger in it."

"You did no harm in that, John ; but turn him round to the light, so that I may see his face."

This being done, I lifted the flap of his nostril to see if there was any indication of inflammation or redness, when I discovered traces of linseed oil within the nose.

My next query seemed to stagger the worthy John, as I inquired if the animal had vomited after receiving the oil (for be it known that a horse only vomits through his nose).

"Vomit," replied John, "what put that in your head, sir ?"

"The oil inside the nostrils, John !"

"Oh ! I see," returned John ; "but we held up his head, and poored the oil doon his nose."

Here, then, was the explanation of the congestion. They had choked the poor animal, a portion of the mixture finding its way into the lungs, which ultimately terminated in the death of a good five-year old horse. Very many, then, are the obsolete remedies still practised. As proof of this, I was in confidence told only the other day by a horse dealer that cow manure and sweet milk used to be administered to roarers, and was a perfect success for the time being.

I have also known cows receiving tar and egg shells; while ground glass blown into the eye is often employed to remove a foreign body from that organ.

Many quack mixtures are still compounded and administered, which are diametrically opposed to ordinary reason, and agents that are antagonistic in character are thrown together and given to avert certain complaints. External remedies are also often indiscriminately applied, the ulterior consequences frequently ending in death, or perpetual blemishment.

Proceeding then to discuss the true remedies that are applicable to domestic animals, we shall notice, in the first place, that pre-eminent agent,

ACID CARBOLIC:

Its Properties, Action, and Uses.

There are few medicines in the British Pharmacopœia which demand greater attention than carbolic acid. To the agriculturist a knowledge of this drug is of the greatest importance, nor should this knowledge be confined to the, farmer. It ought to be disseminated widely, because it stands prominently forward as the best disinfectant agent that the world has yet seen; and were it for no other property than this it cannot be too highly extolled.

It was discovered in the year 1843, as an ordinary constituent of the urine of herbivora, and one of the numerous interesting products obtained from coal tar,—the commercial supply being got from the black and heavier coal tar oils.

Carbolic acid is a narcotic irritant poison. It is used in medicine as an antiseptic and stimulant, and possesses inestimable qualifications as a disinfectant.

A strong solution applied to the skin or mucous surfaces

acts as a topical irritant, coagulates albumen, and leaves a white, dry roughened surface, more or less deprived of sensation, from which the shrivelled outer skin subsequently peels off. When it is swallowed, besides exciting this local action, and also usually causing vomiting, it is readily absorbed in all probability as a carbolate, and produces inebriant effects which are in every way analogous to those of alcohol.

Like alcohol and oil of turpentine, it exercises its influence over the great nerve centres, notably the spinal chord, stimulating, deranging, and paralysing its functions, thereby destroying life by respiratory arrest.

Poisoning is accomplished by whatever channel the acid is introduced into the system. Dressings employed in human surgery sometimes produce nausea, vomiting, giddiness, &c. Again, sheep that are suffering from scab, if too freely dressed, will often have an attack of congested or inflamed lungs, which will in all likelihood terminate life at the end of a few weeks. Dogs are also extremely susceptible to its influence, and one incautious application will produce dulness, depression, trembling, and disinclination for food of the most tempting nature for several days. Stronger dressings, if too freely used, will within a few minutes cause excitement, panting, unsteady gait, and frequently even fatal coma.

The urine of animals receiving carbolic acid, or to whose skins it has been freely applied, gives off a tar sort of odour when treated with a drop or two of sulphuric acid, while a blue colour is developed on adding to it a little iron perchloride solution.

Animals dying from the administration of carbolic acid present the following noteworthy *post-mortem* appearances. There is whitening of the mouth and gullet, and frequently the stomach presents the same sign. Strong solutions leave patches of redness, and inflammation in the stomach and small intestines, while a smoky sort of creasote odour pervades the body. The internal organs are congested, and the vessels of the brain are full of fluid blood.

Now it must be remembered that the great value and manifold application of carbolic acid depends upon its antiseptic and disinfectant power. It arrests as well as prevents fermentation. Yeast treated with it fails entirely to exert its familiar effects. In its presence mould and fungi cannot grow, and if already in existence they are effectually and speedily killed. Nor does it require large doses to effect this, as $1\frac{1}{2}$ per cent. can destroy all organised life in putrefying solutions, putrefactive taint, and fungi as well. Meat steeped for one hour in a solution of 1 per cent. is effectually preserved ; and again, not only does carbolic destroy the lowly cells or germs, which excite fermentation and decay, but oats, barley, and beans, when soaked in a 1 per cent. solution, do not germinate, while plants watered with it die. The vapour or solution promptly poisons not only simple animalcules, but also creatures higher in the scale of life, such as fleas, moths, ticks, earth worms, ascarides, and lumbrici ; in fact, all the lower forms of life are quickly destroyed by solutions containing one part of carbolic acid to one hundred of water.

Again, experiments have proved that the virus of cattle plague loses its reproductive powers when exposed to carbolic vapours. The subtle actively reproducing germs which are given off during the progress of other catching diseases, are doubtless also destroyed when brought into contact with this agent. In short, no substance proves so generally serviceable as a disinfectant as carbolic acid.

We pass on now to the consideration of its medicinal merits. Now, the administration of carbolic acid is clearly indicated in those diseases where tissue change is unduly violent. It has proved of immense value in cattle plague, as it possesses the power of lowering advancing temperature, and prolonging where it did not actually save life. In the Texas cattle fever the remedy most relied on was 12 ounces of carbolic acid, 12 ounces of sodium bicarbonate mixed with 4 ounces of glycerine,—the dose being from one to two tablespoonfuls three times a day in a quart of water. It also mitigates the severity of foot-and-mouth disease, and when given in this and other contagious affections it probably checks the reproduction of the specific virus, and thus prevents, or at all events greatly reduces, its risk of spreading. As an antiseptic in surgery, it has saved both man and the lower animals an enormous amount of suffering. It has prevented blemishing, conserved useful members, and expedited cure. For general purposes no antiseptic is so convenient and so eminently effectual as carbolic acid. Wounds, whether inflicted by accident or the operator's knife, when at once freely saturated with the carbolic solution, and kept covered with lint or other appliances well soaked with the antiseptic, heal rapidly with the best results. All exudates from serious wounds immediately irrigated with carbolic acid solutions, and protected from the air, have no tendency to putrefy or develop foul discharges, which are ever common in wounds. Therefore not only is the health and comfort promoted, but annoyance and injury to other animals in the immediate vicinity is also prevented. Few remedies more rapidly remove the irritating itching pain and swelling occasioned by the stings of wasps, bees, &c. ; while inhalation either with air or steam is of great service in malignant sore throat and unhealthy strangles, abscesses, &c. Considerable care, however, must be exercised either by inhalation, injection, or even as an external dressing, as it is liable to become absorbed, and in this way it develops its narcotic irritant properties.

As an antiseptic and disinfectant, nothing better can be employed for the purification of stables, cow houses, piggeries, kennels, and poultry pens, or for railway trucks, horse boxes, &c.

To ensure thorough purification of infected places, the antiseptic must be often and freely employed, either in the condition of vapour, spray, fluid, or powder, or in several of these forms. The vapour is easily procured by sprinkling the acid on red hot coals, or upon a hot metal plate. In large buildings, besides smearing the walls and woodwork, sheets saturated with it should be hung up here and there, as these catch the floating particles of contagion. The daily use of this agent internally and externally will prevent with tolerable certainty healthy animals from becoming affected, although they are in close proximity to the diseased ones ; and no antiseptic is so effectual in preserving hides, skins, or wool.

Doses.—Horses and cattle take from 15 to 40 drops ; sheep and pigs from 5 to 8 drops ; dogs, 1 to 2 drops.

The crystallised acid dissolved is by far the best for internal use. It is often administered in the form of a ball, but it is better given with glycerine, and is not likely then to produce local irritation. For external purposes, it is often mixed with soap, the preparation being from 1 to 12 or 16 as an application for ticks, lice, mange, ringworm, &c., and for wounds with oil in the same proportion.

The next disinfectant on the list we must assign second place to, is known as

CHLORIDE OF LIME or BLEACHING POWDER.

This agent was first introduced and prepared by Messrs Tennant & Mackintosh of Glasgow, in the year 1798, and is a valuable bleaching agent. In fact, most of the laundries now-a-days resort to it in order to save labour and whiten the calico. It is a soft grey white powder, with an odour of chlorine, or to be more accurate in the description, of hypochlorous acid. It possesses an acrid bitter taste, and deliquesces when exposed to the air. It is to a certain degree soluble in water, a portion of the lime remaining insoluble.

It is an irritant, stimulant, alterative, astringent, desiccant, a feeble antiseptic, a particularly good deodoriser, and a disinfectant.

It has been recommended for tympanitis in the horse, and hoven or distention of the stomach in cattle. But repeated trials have now convinced the veterinarian that for these complaints it is of little use. It is used, however, externally with more favourable results, and acts as a stimulant and deodoriser for the following unhealthy wounds, such as canker, thrush, and fistula. Now, although it does not possess the power to arrest putrefaction or prevent fungus growths, it nevertheless attacks and successfully breaks up the products of putrefaction, therefore it is an efficient deodoriser, and a tolerable disinfectant. It is often spread at gateways, cowsheds, and other premises where contagious disease exists. It has a sickly smell, and is as a rule disliked by man and beast. It diminishes the value of manure, and if scattered about a stable or cowshed, it will banish flies; while neither rats nor mice will seek a habitation where it is sprinkled, more especially if it is mixed with sulphur.

Doses.—Horses can take from 1 to 2 drachms, cattle 2 to 4 drachms, sheep about 1 drachm, dogs 2 to 5 grains, to be administered either in cold gruel or milk, or given in a ball.

CAMPHOR.   9

CAMPHOR

Is obtained from a tall, handsome evergreen, which grows in China and Japan. From experiments made by Sir Robert Christison of Edinburgh, it yields about 1-500th of its weight of camphor, which is sometimes extracted by exposing the wood to dry distillation. In Formosa, from which most of the camphor imported into this country comes, the branches are steeped in water and boiled, the fluid is strained and allowed to stand until it concretes, the crude camphor with alternate layers of dry earth, is placed on copper vessels, over which domes are inverted, into which, on the application of heat, the camphor sublimes. Excessive doses are irritant and narcotic. The medicinal properties—anodyne, stimulant, diaphoretic, and slightly diuretic. Externally it is stimulant, irritant, and feebly antiseptic.

The poisonous doses, two drachms of camphor injected into the jugular vein, or two ounces introduced into the stomach, will kill a horse. Hertwig considers the deadly dose for a dog to be from a quarter to half an ounce.

Symptoms of poisoning by this agent.—General excitement, affection of the mucous membranes, muscular twitchings, vertigo, profuse sweating. The odour of camphor can be detected in the breath and the secretions of the body. The animal falls to the ground with outspread nostrils and dilated pupils, and foaming at the mouth. Smaller animals are frequently affected with partial paralysis and loss of sensibility, often apoplexy, with loss of motion and sensation, and death ensues without convulsions.

Post-mortem appearances.—Although classed with narcotic poisons, it nevertheless produces irritation of the alimentary canal. The blood is dark, and coagulated in the heart and larger vessels. The body exhales a strong odour of camphor, which is quite perceptible even after the flesh has been boiled.

Treatment.—Acidulated draughts and purgatives. If the dose does not prove fatal, diuretics assist in expelling the drug by the kidneys. In small animals emetics must be given.

Doses for horses 1 drachm, cattle 2 to 4 drachms, sheep and pigs 22 to 40 grains, and for dogs 5 to 10 grains. When given internally, it is best made into an emulsion with eggs, or dissolved in milk or oil. For external use, it ought to be dissolved in six or eight parts of alcohol, dilute acetic acid, oil of turpentine, or linseed oil.

CANTHARIDES or BLISTERING FLY.

These flies are found in Southern Europe, Germany, and Russia, and sometimes along the south coast of England. They generally take up their habitation upon such trees as the olive, lilac, ash, alder, honeysuckle, and rose. During May and June, after nightfall or before dawn, the collectors, with their faces and hands protected, shake or beat the insects from the trees on which they feed, and kill them by exposure to the fumes of the oil of turpentine, or by immersion in boiling vinegar, and quickly dry them either in the sun or by artificial heat. Most of the flies used in this country were formerly brought from Spain, hence the name Spanish fly.

The poisonous doses of this insect are half an ounce and upwards for horse or ox; a drachm for a sheep, and half a drachm for a dog.

Symptoms of Poisoning.—The mouth and fauces, and indeed the whole of the alimentary canal, is excessively irritated or inflamed, the membrane of the mouth is red, and there is considerable difficulty in swallowing. Dogs vomit, and efforts of a similiar nature are made by the horse and ox. The urinary organs are very much affected. In all animals there appears to be a certain degree of sexual excitement. The urine is voided with difficulty and pain, as it is irritating and burning ; it is highly albuminous ; the fæces are covered with mucus, and mixed with blood. At first the animal appears excited, but very soon prostration of strength supervenes, cold sweats bedew the body, there are muscular twitchings, paralysis of the hind quarters, and finally death.

Post-Mortem Appearances.—The intestinal canal and urinary organs are congested or inflamed. The kidneys may appear healthy, but the lining membrane of the bladder is always reddened. There are black patches and even spots of ulceration in the intestines.

Treatment.—Mucilaginous and albuminous drinks, for example linseed tea, a strong emulsion of gum arabic, and the white of eggs. Oil is to be avoided, as it dissolves the cantharidine, and favours its absorption.

Cantharides is chiefly used in the form of powder, oint-ment, tincture, liniment, and blisters, and as such possesses great value. Given internally, the dose for a horse is from 4 to 20 grains, for cattle 10 to 20 grains, for sheep and swine 2 to 8 grains, for dogs half a grain to 2 repeated once or twice a day, usually administered with aromatics and bitters, and should any untoward effects occur it must be suspended at once. The tincture of cantharides is vulgarly termed a sweating blister.

AMMONIA or SPIRITS OF HARTSHORN

Exists in the air, in dew and rain, in some mineral springs, and in the juices of most plants furnishing their nitrogen.

It is also freely evolved from the putrefaction and destructive distillation of organic matters, and from the bodies of living animals. But the coal beds are the great commercial source of ammonia and its compounds. Ammonia in its gaseous state is irritant; if respired it inflames the throat and bronchial tubes in a violent manner; in concentrated solution it is an irritant caustic poison; in medicinal doses it is stimulant, antispasmodic, antacid, antiseptic, diaphoretic, and diuretic; applied externally, it is irritant and vesicant.

Hertwig discovered that half an ounce of the strong solution had no bad effect on horses, but one ounce proved fatal in sixteen hours, and three ounces in fifty minutes, the latter quantity causing violent cramps and difficulty in breathing. Half a drachm introduced into the stomach and secured by tying the gullet, destroyed a dog in twenty-four hours, producing much uneasiness, agitation, and stupor, and leaving after death slight redness of the villous coat of the stomach. When injected into the veins, it causes spasms, convulsions, and death, which as a rule occurs in a few minutes, and depends upon the sudden arrestation of the heart's action. The most effectual antidotes are vinegar and other diluted acids, with diluents and demulcents.

Its Medicinal Properties.—Ammonia is beneficial in influenza, scarlatina, purpura, chronic bronchitis, and pneumonia—wherever indeed there is failure of the heart's action, cold extremities, or sudden depression of the vital powers. It relieves congestion of the lungs brought on by over-exertion or exposure. For all such purposes it may be both inhaled and swallowed. When given by the mouth it is usually combined with alcoholic stimulants. In chronic bronchitis, particularly in horses, ammonia inhalations, besides acting as general stimulants, lessen over-abundant secretion, but great care must be taken that the pungent gas is not used too freely or too concentrated.

In indigestion, hoven, and tympanitis, its twofold action as a stimulant and antacid render it especially useful. Controlling irregular or inordinate nervous force, it counteracts the spasm of colic and epilepsy, and in the latter disease may be given by the mouth as well as cautiously inhaled. It is a valuable antidote in poisoning by prussic acid, opium, tobacco, and other sedatives. A few drops are sometimes added to emetics to increase their activity and lessen their depressing influence. It is used both internally and externally in the treatment of snake bites, but its success is uncertain. Diluted solutions abate the irritation induced by stings of wasps. Mixed with oil or other bland fluid, it is applied as a counter-irritant in sore throat, bronchitis, rheumatism, and chronic disease of the joints in cattle. As an external stimulant, it acts very rapidly, has no tendency to irritate the urinary organs, and is used either alone or in conjunction with cantharides, mustard, or turpentine. It is an effective antiseptic.

Doses.—As a diffusible stimulant, horses can take from 2 to 6 drachms, cattle 2 to 10 drachms, sheep and pigs 1 drachm, dogs from 5 to 12 drops. The medicinal solution being half the strength, is given in double the above quantities. They require to be repeated every two or three hours. A useful stimulating draught, either for horses or cattle consists of the following:—Half an ounce of medicinal ammonia, half an ounce of sweet spirits of nitre, half an ounce of the tincture of gentian, to be given in a quart of ale.

CARBONATE OF AMMONIA.

The carbonate resembles closely the ammonia hydrate, but is less irritant. It is sufficiently active, however, to produce in small animals the same irritant poisoning. Two and a half drachms given to a dog caused gastric inflammation, tetanic convulsions, and death. Notwithstanding, for all classes of patients, it is the most convenient ammoniacal compound, and is employed, often advantageously conjoined with alcohol, in influenza, scarlatina, erysipelas, also in the second stages of inflammatory complaints and in convalescence from debilitating diseases. To maintain the flagging vital powers, it is recommended in the puerperal apoplexy of cows, while it frequently averts epileptic fits in weakly dogs.

Doses.—Horses require 2 to 4 drachms, cattle 3 to 6 drachms, sheep and pigs 15 to 40 grains, dogs 3 to 8 grains. It is given either in a ball with linseed meal, or better still dissolved in gruel, which ought to be cold. Where prompt stimulating effects are required, it is frequently conjoined with alcohol, ether, or sweet spirits of nitre ; whilst in chronic ailments it is advantageously united with gentian, ginger, oak bark, and other tonics. The best smelling salts are made by adding to the sesquicarbonate half its weight of strong ammonia solution, adding some aromatic oil, such as bergamot or lavender.

ALCOHOL.

Alcohol is represented in the British Pharmacopœia in several distinct forms, such as absolute alcohol, rectified spirit, proof spirit, and methylated spirit, and is also extensively used in several forms of wine and spirits.

The various alcoholic fluids are obtained either directly or indirectly from the fermentation of saccharine solutions.

Rectified spirit, or spirits of wine, are the usual terms applied to alcohol obtained from the distillation of fermented saccharine fluids, containing 16 per cent. by weight, or 11 per cent. by measure of water, with the specific gravity of ·838. To absolute alcohol it bears a general resemblance, but has less pungency and volatility, and a higher boiling point. It is used for making all spirits and many of the tinctures and extracts that are sold.

The following alcoholic fluids, employed both dietetically and medicinally for man, are also frequently prescribed for the lower animals.

Wine, the fermented juice of the grape, contains from 5 to 17 per cent. of excise alcohol. Brandy, prepared by the distillation of the weaker wines, contains about 53 per cent. of excise alcohol. Rum, a fluid about the same strength, is made by the distillation of a fermented solution of molasses. Whisky is of similar strength, and is obtained by distilling a thoroughly fermented solution of malt, or of malt and raw grain. Whilst Hollands, Geneva, and gin, a little weaker than the foregoing, are prepared from fermented malt, and a small quantity of juniper berries. Ales and porter, convenient stimulants in almost every-day use, are made by infusing malt in water, at about 180°, allowing it to stand for a few hours, until the starch is in great part converted into dextrine and sugar, boiling the solution with the requisite hops, adding yeast to cause fermentation, which must be carefully watched lest it should go too far. The dark colour of the porter is due to a portion of the malt being roasted. Porter and ales contain between 4 and 8 per cent. of excise alcohol.

The Action and Application of Alcohol.—It is a narcotic poison, and belongs to the inebriant class. It first stimulates, then deranges, and ultimately depresses the functions of the brain and spinal chord. It kills as a rule, by paralysis of respiration. Medicinally it is a very valuable diffusible stimulant, antispasmodic, cardiac tonic, and antiseptic. During their excretions from the body, they act as diuretics and diaphoretics. Carefully used in small doses, alcohol is a readily assimilable article of food; locally, it an irritant and refrigerant.

Strong spirit applied to the skin or to a mucous surface produces increased vascularity, accompanied by heat, redness, and irritability; these, however, are soon superseded by diminished vascularity and sensibility. Similar characteristic signs occur when alcohol is taken in full doses internally. The general excitement first observed is followed by deranged and depressed action. It coagulates albumen, and destroys those organic germs which excite fermentation and putrefaction. It exerts similar effects by whatever channel it enters the body, whether injected into the veins, inhaled, or swallowed. Moderate doses given by the mouth increase the gastric secretions and aid digestion, but large doses and strong solutions destroy the pepsin, arrest secretion, irritate and inflame the mucous coat, and thus interfere with absorption.

There has been much controversy, and I suppose will be, as to the dietetic value of alcohol. It contains no nitrogen, and hence it is not available for the building up of the albuminous structures. But like sugar, starch, or fat, it undergoes oxidation or combustion. It supplies force and heat; it maintains or increases the weight of the body, and when used in excess frequently causes an accumulation of fat or fatty degeneration. Men and the lower animals kept on somewhat deficient diet, on which weight would be lost, maintain their weight when receiving in addition daily small doses of alcohol. In health alcohol can easily be dispensed with, and there can be no doubt whatever that those systems are the most robust (and better able to withstand disease) that never taste it. But in fever the retardation of oxidation adds greatly to its value, because it checks undue tissue waste, it lowers the temperature, while it is more readily digested and assimilated than ordinary food. In short, it is a valuable servant, but a despotic tyrant when it becomes master, because its commandment is "Thou shalt have no other God but me."

Its Medicinal Properties.—I do not know of another drug that is more extensively used than alcohol. It possesses the most extraordinary power that the mind is capable of conceiving. It will rouse and regulate weak and disturbed nervous power; therefore it is useful in indigestion, it is useful in spasmodic colic, it is useful in arresting chills, it is useful in cases of poisoning by either aconite or tobacco. It strengthens the action of the heart, dilates the capillaries, and by doing so it diminishes the arterial tension, and in conjunction with heat it increases animal temperature. For these properties it deserves a word of commendation. It is valuable in influenza, in debilitating disorders, and in convalescence from acute attacks. Under its powerful sway the appetite improves, the temperature becomes normal, and the hurried pulse becomes firm and steady. Where a horse is hard wrought, with a pulse quick and weak, with the breathing hurried and embarrassed, no treatment in the world is better or more successful than a couple of ounces of spirits every two hours, and mustard to the throat and sides. Therefore, I must again repeat that it is dangerous to condemn an agent with such valuable inherent properties.

Doses.—The rectified spirit, to horses 1 ounce, to cattle 1 to 3 ounces, sheep half an ounce, and dogs 1 drachm.

Whisky, ale, brandy, gin, wine, &c., according to the patient's condition; but small doses, oft repeated, are best in all circumstances.

BELLADONNA OR DEADLY NIGHTSHADE

Grows wild in most parts of Great Britain, more especially about old walls, the edges of plantations, and shady ruinous places. But the demand having increased, it is now carefully cultivated. It is an annual stem, and grows from three to five feet high; the leaves are four or five inches long; the flowers are pendulous and bell-shaped, and of a dark purple tint.

Its Medicinal Properties.—It is decidedly useful in influenza, tetanus, scarlatina, and purpura in the horse; also in sore throat, inflammation of the lungs, bronchitis, and diseases of the heart. It will arrest congestion; it soothes and pacifies the nervous system; while to an animal suffering from inflammation of any of the organs of the throat, no other remedy gives such quick and prompt relief. Given with a stimulant, such as the carbonate of ammonia, it will relieve without delay the painful cough which so often occurs in distemper in dogs. In inflammation of the bowels, in rheumatism, in epilepsy, and chorea, its services are very desirable.

Applied externally, it relieves irritable and painful wounds, and the raw surface of frost-bite, cracked heels, and mud fever; given in the form of an injection, it allays irritation of the bladder and rectum, and counteracts spasmodic contraction of the womb.

Inflammation of almost any part of the eye is allayed by the application of belladonna.

Doses of the Extract.—Horses 1 to 2 drachms, cattle 2 to 3 drachms, sheep 20 to 30 grains, dogs 2 to 5 grains.

OPIUM

Is obtained from the poppy species. Its narcotic properties are due to the morphia, with many other alkaloids, in very variable quantities, combined with sulphuric acid and a peculiar organic acid, the meconic.

Its Action and Uses.—Containing as it does so many different constituents, its action of course must necessarily be diversified; it acts on the spinal and brain systems, and also upon the sympathetic. A full dose has the same effect as alcohol—stimulant first and a depressor afterwards. In horses the excitant action predominates, in dogs the two antagonising actions are more evenly balanced; there is as a rule, usually delirium with stupor. Poisonous doses prove fatal by arresting the respiration; applied externally, it first stimulates, then soothes and paralyses both sets of nerves. Therefore, whether employed internally or externally, it is one of the most effectual antidotes for pain, nervous irritability, and spasms that we possess.

Now horses like men display considerable difference in their susceptibility to the action of this medicine; excitable well bred animals are brought under its excitant effects far more readily than a coarse bred one; therefore to this fact must be ascribed the many contradictory reports regarding its action.

The ox and cow are not very susceptible to the influence of opium; a cow will take an ounce, and a sheep 4 drachms, without any other effect than dryness of the mouth. Swine after they get 1 or 2 drachms become lively, then dull and sleepy, their bowels get constipated, and their skin hot.

Dogs manifest the same symptoms as man under its influence. With moderate doses most people become stupid and drowsy, but other individuals are rendered delirious. In animals poisoned by opium, the blood is fluid and dark coloured, from imperfect decarbonisation. There is general venous engorgement. There is no medicine more frequently prescribed. As a stimulant and restorative, it acts almost like food or alcohol. The Cutchie horsemen share their opium with their jaded steeds, and increased activity and endurance are the effects both in man and beast.

It is of great value in colic, in inflammation of the bowels, in peritonitis, and all inflammatory affections situated within the abdomen, because it combats spasms, irritation, and pain.

Opium is best administered in the form of the tincture or laudanum. The immediate effects are more apparent. The dose of the fluid being for horses and cattle from 1 to 3 ounces, for sheep and pigs 2 to 4 drachms, for dogs 15 to 40 drops. In diarrhœa and dysentery, there are very few remedies more effectual than injections of the tincture of opium or laudanum, mixed with warm starch gruel.

ACONITE or MONKSHOOD.

Botanists have counted twenty-two species and upwards of one hundred varieties of aconite. Some of these species are, however, inert or very nearly so. According to authority, the root is six times as active as the other parts, and is consequently most valued. The leaves are less active than the root, but more so than the flowers, fruit, or stem.

Its Action and Uses.—It paralyses both sets of nerves, that is, the sensory and motor. It causes spasm of the muscles of the chest, arrests respiration, when death ensues from suffocation. As a sedative, it is safer and easier managed than bleeding; it is more certain and effectual than either calomel, opium, or tartar emetic. When chewed or rubbed on the mucous surfaces or skin, it produces a peculiar tingling sensation with numbness; there is no alteration of structure, there is no irritation or excitement produced, the sensory nerves appear to have lost all feeling, and seem paralysed. This agent seems to possess a uniform effect upon all animals, even from the earthworm to man. Horses receiving an overdose tremble violently; they lose the power of supporting themselves, become slightly convulsed, froth at the mouth, perspire freely, appear much nauseated, and attempt to vomit, the breathing becomes slower and feebler, the pulse is reduced in strength and in number, and it will require six or eight hours before the breathing and pulse regain their normal standard; impared appetite and sickness will remain for a couple of days.

Its Medicinal Properties.—Aconite is a prompt and effectual sedative for all domestic animals. In a quarter of an hour after a dose is administered, the number of pulse beats is reduced one-fourth, their force is weakened, vascular excitement is thus abated, the high temperature is lowered, and the pain is relieved. No sedative is so certain and successful in the early stages of inflammation of the lungs, pleurisy, bronchitis, enteritis, peritonitis, laminitis, or rheumatism. Indeed, it is certainly the most effectual and valuable agent that the veterinary surgeon possesses for controlling at the commencement attacks of acute inflammation or fever, either in horses or cattle. If a horse has been hard-worked and exposed to a chill, accompanied with acute sore throat, a little mustard externally, and a couple of doses of aconite is generally all that is required. Shepherds employ it with great success during the lambing season upon the ewes.

Fleming's tincture is the ingredient most employed upon our patients, as it is about four times the strength of the B. P.

Dose.—For a horse 10 drops, for cattle 10 to 20, for sheep 2 to 3 drops, and for dogs 1 to 2, given in water or gruel, and repeated every two hours.

CHLOROFORM.

This medicine was discovered in 1832 by Soubeiran and Liebig. Its effect upon the lower animals was described by Dr Glover in 1842; while its valuable anæsthetic action was first discovered and applied by the late Sir James Y. Simpson of Edinburgh, in November 1847. Since that period it has been extensively employed for the mitigation of human suffering during surgical operations, parturition, and numerous diseases. It has also crept into our practice for similar purposes. The British Pharmacopœia gives the following explicit directions for making and purifying:—

"Mix 30 fluid ounces of rectified spirit with 3 gallons of distilled water in a capacious still. Add 10 pounds chlorinated lime, thoroughly mixed with 5 pounds of slaked lime. Let the condenser terminate in a narrow-necked receiver, and apply heat so as to cause distillation, taking care to withdraw the fire the moment the process is well established. When the distilled product measures 50 ounces, remove the receiver, and pour its contents into a gallon bottle half filled with water, shake well together, and set at rest for a few minutes, when the chloroform will subside; pour off the water, and thrice wash the chloroform in a smaller vessel with successive portions of 3 ounces of water. Agitate the washed chloroform for five minutes with an equal volume of sulphuric acid, then, after subsidence of the latter, transfer the chloroform to a flask containing 2 ounces of chloride of calcium in small fragments, mixed with half an ounce of perfectly slaked lime. Mix well by agitation. After the lapse of an hour, connect the flask with a Liebig's condenser, and distil the pure chloroform by means of a water bath. Preserve the product in a cool place, in a well-stoppered bottle. The lighter liquid which floats on the crude chloroform after its agitation with water, and the washings with distilled water, should be preserved and employed in a subsequent operation."

Chloroform is a transparent, colourless, neutral, oily-looking, mobile fluid, with a sweet taste, and a fragrant, ethereal, and apple-like odour.

Its Actions and Uses.—Full doses paralyse and narcotise the brain and spinal systems, by whatever channel it enters the body. Death results from respiratory arrest and suffocation. The vapour inhaled speedily induces anæsthesia. Taken internally, it is stimulant, antispasmodic, and anodyne. Undiluted, it is a topical irritant. It is applied externally as a rubefacient, anodyne, and local anæsthetic.

An ounce rapidly swallowed by a dog 15 lbs. weight, or half an ounce injected into the chest, causes a sudden cry, gasping respiration, a run of feeble pulsation, and death in 70 or 80 seconds. Similar rapid and fatal effects result when small animals are introduced into a large glass jar containing 7 per cent. of vapour ; it rapidly enters the blood, but the changes it produces there are unknown at present. It is removed from the body mostly by the lungs, but in smaller amount by the skin and kidneys. It acts directly on the brain and nerve centres, paralysing and totally extinguishing their functions. The action of the heart at first is quickened then depressed. The pulse eventually becomes rapid and weak, the respiration at first retarded, now becomes quicker; but as narcosis increases, it gets more slow, shallow, and irregular.

The pupils are at first contracted, they then gradually dilate. In dogs, cats, and rabbits, as in the human subject, apparently from paralysis of the sympathetic nerves, the functions of the liver are deranged, and sugar is found in the urine. In these and other effects of chloroform functional activity is succeeded by functional paralysis. Sir James Y. Simpson considered that chloroform saves the lives of six persons in every hundred subjected to surgical operations; therefore, with proper precaution, it is a safe remedy. It has been administered many thousand times in Scotland and in England too. During the Crimean war 30,000 French soldiers inhaled it without a single accident; while on our side, with 50,000 men under its influence, there were only two deaths. In the American war it was given 22,000 times without accident; therefore the various deaths do not average more than 1 in 17,000. I have employed it successfully on both the horse and cow. In the case of a cow she never could be induced to retain the seminal fluid after being served by the bull. Many and frequent were the attempts made by rubbing her back, douching her with cold water, &c.; still she always succeeded in voiding it. I was requested to call one morning when she was to be served. I took with me four ounces of chloroform, and as soon as the process was complete I placed the nose-bag over her mouth, tied it over the head, and allowed it to remain until she was completely under its influence. She rested in a comatosed state for half a day, and pregnancy was the result. In the horse case his disorder consisted in his extraordinary viciousness when taken to the forge; he did not only damage himself, but men were in danger of their lives. He was a valuable hunter, so we did not want him blemished. We had no stocks, therefore I had to resort to chloroform, and I may here say with the most satisfactory results; when he recovered sufficiently so as to be able to stand, he was quietly shod all round, with the docility of a good-natured old one. I have also administered it with success during parturition in the mare, at Charles Gow's, Esq. of Melvine Hall, Ormiston, N.B.

Doses.—For horses and cattle from 2 to 4 ounces, 1 to 2 ounces for sheep and pigs, and 2 to 4 drachms for the dog. The inhalation is the most simple and advisable way of giving it. Place a sponge saturated with it in a nose-bag, which must be perforated with holes in order to allow the air to enter. If undue effects be produced, the inhalation must be stopped at once, free access to fresh air allowed, throw water over the head and neck, which of course must be cold, and apply artificial respiration.

Chlorodyne.—Now a popular anodyne in human medicine is made from different formulæ. That of Dr Collis Browne's is said to contain ten parts each of chloroform, ether, Indian hemp and morphine, two parts capsicum tincture and prussic acid, three parts aconite and hyoscyamus tinctures, one part oil of peppermint, five parts of hydrochloric acid, and fifty of simple syrup.

CROTON OIL AND CROTON SEEDS.

The tree grows from 15 to 20 feet in height on the Indian continent, in Ceylon, and in many islands. Its nuts are larger than a hazel, of an oval-triangular form, and contains three seeds about the size of French beans, each one weighing about three grains.

They are brown, with no odour, and taste at first mild and mucilaginous, but very soon they become hot and acrid. The seed contains from 50 to 60 per cent. of a fixed oil, which when separated constitutes the croton oil.

The acrid oil thus expressed from croton seeds is an energetic poison. One drachm of the bruised seed will sometimes kill a horse in from five to six days. Two drachms give rise to great fever, colic, general debility, and in from six to fifteen hours superpurgation sets in. The pulse cannot be felt at the jaw, cold sweats bedew the body, and death takes place in about twenty hours, and if the animal is weak its life will terminate in ten hours.

From 10 to 20 grains excite violent purgation in the dog ; and if the gullet be tied the smaller dose will induce efforts to vomit, paralysis, and death in four hours. Both in the horse and dog, after death the stomach and intestines are found inflamed ; frequently there are erosions of the mucous membrane and effusions of blood in the intestines, and in some cases the lungs are inflamed.

A drachm of croton introduced into the cellular tissue of a dog's limb, brought about complete loss of sensation, and arrested the power to move in twenty-eight hours, and after thirty hours produced death.

Hertwig states that eight drops injected into the jugular vein killed a horse, while two drops applied in the same manner killed a dog.

Its Medicinal Action and Uses.—It is employed as an active purge—for cattle suffering from impaction, fardel-bound, and constipation from torpidity of the bowels and parturient apoplexy, milk fever, &c. It is servicable where you cannot give bulky medicines, when animals are unmanageable or where there is difficulty in swallowing. The bad effects of overdoses is combated by demulcents, and opium given by the mouth and rectum, hot fomentations to the belly, and afterwards stimulants to counteract depression.

Doses.—Horses take 36 grains of the seed, therefore, if you allow 3 grains to each seed, you will require 12 seeds. 20 seeds are given to cattle, 4 for sheep, 3 for pigs, and 1 or 2 for dogs. The dose of the oil is, for the horse 15 to 25 drops, for cattle ½ to 2 drachms, for sheep and swine 5 to 10 drops, and for the dog 2 to 3 drops.

If you add the oil to any blistering composition it greatly increases its irritable properties.

DIGITALIS or FOXGLOVE

Grows wild in this climate, and in numerous parts of the Continent, in young plantations, by the hedge sides, and hilly pastures. As a medicine, this agent has been highly praised, owing to its decided and marked influence upon the heart. It is, however, capable of irritating the alimentary canal, is sedative to the nerve centres, and through them affects the heart.

According to Delafond, the poisonous dose for the horse is from 1½ to 2 ounces; and Hertwig says that 6 drachms are sufficient to produce symptons of poisoning; half an ounce may induce inflammation of the stomach and bowels. Cattle suffer from large doses, and dogs die from the effects of from 2 to 3 ounces, after it has been administered—say from six to eight hours ; providing it is a poisonous dose, there will be general dulness, loss of appetite, staring coat, inflamed mucous membranes, staring prominent eyes, dilated nostrils, breathing and pulse hurried. In the course of twelve hours intestinal irritation, with sickness, colicky pains, purging, and in some animals vomiting. But the characteristic symptom of poisoning by this digitalis is, the violent and intermittent action of the heart, the pulse feeble and indistinct, respiration fast, then becoming slow, irregular, and interrupted, rapid emaciation of the body, at first deficient urinary secretion, spasmodic efforts of the bladder, and lastly, copious and profuse flow of urine.

The Post-mortem Appearances reveal inflammation of the stomach and bowels, black uncoagulable blood, the ventricles of the heart contracted, and the auricles dilated. Digitalis is one of those agents that accumulate in the system, and for some time without any apparent effect ; but it may begin abruptly to act with great energy, as if with the accumulated power of all that has been administered, and symptoms of poisoning are manifested.

In treating cases of poisoning by this agent, substances should be given that contain tannin, as this renders the poison inert and insoluble. If the system is much prostrated, then stimulants must be given in order to support life until the poison is expelled.

Doses.—Horses take of the powdered leaves 10 to 30 grains, cattle ½ to 1 drachm, sheep 8 to 15 grains, pigs 2 to 10 grains, dogs 1 to 2 grains. Owing to its effects being prolonged, it should only be given once in twenty-four hours; but half doses may be given night and morning. Stonehenge advises for a medium-sized dog half a grain of digitalis, nitre 5 grains, ginger 3 grains, made into a pill with syrup, or linseed flour and water.

ERGOT OF RYE.

This is a disease found in all grasses, but principally in rye. Hence the appellation, ergot of rye. It abounds in many countries. Ergot of rye is highly valued as a uterine excitant; but, if eaten regularly, is one of the most horrible poisons known, and causes mortification of the limbs. The ergot of maize is, according to Roulin, very common in Columbia, and the use of it is attended with a shedding of the hair, and even the teeth of both man and beast. Mules fed on it lose their hoofs, and fowls lay eggs without a shell.

The condition induced when an animal partakes of ergot for some time is called ergotism. One large dose produces in man and animals dryness and irritation of the throat, salivation, thirst, burning pain in the stomach, vomiting, colic, and sometimes diarrhœa, brain symptoms, such as giddiness, headache, and stupor.

The chronic effects have been observed on birds, and pigs, and dogs. The first effect is to produce loss of appetite and stupefaction. When it begins to act dogs howl in a frightful manner until they are completely under its influence, they then lie down and groan.

Symptoms.—Dull stupid expression, staring look, dilated pupils, vertigo, signs of inebriation, coma, tremors, convulsive twitchings, tetanic spasms especially of the hind legs, which soon become feeble and paralysed; the animal can scarcely stand, moves slowly, and with great difficulty. There is general debility and loss of flesh, pulse slow and weak, skin cold, coat staring. The extremities, ears, horns, tail, and legs have lost their natural temperature; there is a sero-mucus or sometimes bloody discharge from the nose, the limbs are swollen, black spots and livid patches break out on the surface of the body, dry gangrene of the beak and tongue of birds, of the ears, tail, and the joints of the limbs; these parts separate slowly without pain, the dead from the living tissue adjoining.

Post-mortem Appearances, in cases of poisoning by ergot, are more or less irritation of the alimentary canal, the viscera are flaccid and softened, the muscles semi-gelatinous, the blood fluid, and the interior of the heart and blood vessels ecchymosed and red, as in putrid disorders.

Action and Uses in Medicine.—As an agent to excite parturition, it is not often employed in the lower animals; and when it is required, it is in cases where the throes are languid and a long interval between them, where the animal has been in labour for a considerable time, where no obstruction is present, and where the entrance to the womb is considerably dilated. It is unsuitable when there is a malformation either of the mother or the fœtus. It is sometimes prescribed to get rid of cysts, and hasten the expulsion of the placental membrane, or, as it is commonly called, the

cleansing. In conjunction with ice-bags or other cold and
styptic applications, it is administered to constringe bleeding
vessels in uterine and other hæmorrhage.

Doses.—The mare and cow take from ½ to 1 ounce; for
sheep, swine, and bitches about 1 drachm. These doses are
repeated every half hour or hour, and ought to be given in
the state of a watery infusion, tincture, or liquid.

HEMLOCK.

This is the product of *Conium maculatum,* but several
plants are popularly included under the name.

The extract of the leaves and roots of the common or
spotted hemlock have been found by Dr Christison of
Edinburgh to produce paralysis of the voluntary muscles,
with occasional slight convulsions, then paralysis of the
muscles of respiration, and lastly death from apnœa; the
heart continuing to contract long after respiration had
ceased, sensation did not seem to have been impaired at all.

Now, opposed to these results, are the observations of
another authority, who found that in cats, doses, even not
large enough to be poisonous, caused great languor and
drowsiness, and often profound sleep for two or three hours,
the muscular excitability being reduced, and the circulation
and general temperature lessened. After death the appearances are general venous congestion, a fluid state of the
blood, and a softening of the brain.

Ducks have been seen stupefied and paralysed from
eating the seeds of the plant. Milk and oil will save seven
out of nine affected by the poison.

In cattle, the treatment consists in emptying the stomach;
this holds good with sheep also; in the dog he must be
induced to vomit, and in all animals the strength must be

supported by stimulants, and artificial respiration resorted to should the breathing have ceased.

Analysis.—Hemlock yields a volatile oil upon distillation with water, which appears destitute of noxious properties. The active principle is a peculiar volatile alkaloid, *conia*, which exists in the plant, combined with an acid, probably the *coneic*, by which it becomes fixed, so that it is not given over with water in distillation. Conia is a yellowish liquid of oleaginous aspect, strong, penetrating, mice-like odour, and very acrid, benumbing taste. This mouse-like odour can easily be perceived, when the leaves of hemlock are triturated with a solution of caustic potash in a mortar, and affords a splendid test. Conia is an energetic poison, its effects being analogous to those of hemlock. It is obtained by distilling the plant with caustic potash.

There are very few more soothing or pain-relieving mixtures than hemlock, opium, and chloral hydrate. Of the tincture of hemlock the dose in use is for horses and cattle 2 to 3 ounces, for dogs 2 to 3 drachms. Should the extract be employed, the dose is for the larger animals 1 to 2 drachms, and for the smaller animals 1 to 5 grains.

IODINE

Is principally found in sea water. It is extensively prepared in Glasgow by breaking kelp (the semi-vitrified ashes of sea-weeds) into small pieces, and dissolving in water, when sodium chloride, carbonate and sulphate, with potassium chloride, crystallise out. It is also obtained from the mother waters of the Peruvian saltpetre mines in large quantities and perfectly pure.

Its Action and Uses.—Given in large doses, it is irritant and corrosive. Inhalation of the vapour causes irritation,

cough, and spasm of the throat. Medicinally, it stimulates
the secreting glands and vessels ; it is an alterative, and
will arrest thirst and excessive secretion of urine. If continued for a lengthened period, it produces much waste of
the tissues, causing intense debility, with a depraved state,
termed iodism. Externally, it is used as a stimulant and
antiseptic and counter irritant. It is an effectual deodoriser
and disinfectant. Applied to the skin or mucous surfaces,
it produces an orange-yellow stain, with redness and irritation ; placed in the areolar textures, it induces inflammation
and abscesses ; inhaled as a vapour, it excites cough and
bronchial irritation. It is prescribed as an alterative and
resolvent in the second stages of inflammation, after the acute
symptoms are subdued. It assists in the removal of water
in the chest and dropsy, of recent exudations on mucous
membranes and of glandular enlargements. Among the
lower animals it is given in enlargements of the liver and
udder, in rheumatism, more especially of a chronic nature ;
and amongst cattle in mesenteric disease, in pulmonary consumption, enlarged joints, and other scrofulous affections;
and in diabetes and polyuria it is of inestimable value in
the horse. It is occasionally applied in sore throat ; and
rubbed into the chest in pleurisy, it arrests the formation of
the exudate, and hastens the removal of what may have
been poured out. It is useful in skin diseases, such as scab,
mange, and ringworm, often being mixed with sulphur or
mercurial preparations.

Doses.—Horses take 20 grains to 1 drachm, cattle ½
to 1½ drachm, sheep 15 to 40 grains, pigs 10 to 20 grains,
dogs 3 to 8 grains. Such doses may be given once or
twice daily, continued for a week, then withheld for a
couple of days and resumed again.

BROMINE.

This ingredient's properties are analogous in every way to those of iodine. It is a red liquid, and is identified by its colour and odour. It is soluble in water, alcohol, and ether. If it exists as a hydrobromic acid or a bromide, chlorine should be passed through the suspected liquid. A red colour results from the bromine being set free, which may afterwards be separated by ether.

IRON.

We need not occupy space by giving a detailed account of the manufacture of iron, for most people are familiar with the process. We will, therefore, confine ourselves to its action and uses. It appears, then, from chronicles handed down to us that it was one of the first mineral substances employed as a medicine, and some of its compounds have been administered for 3000 years. As filings, it is occasionally given in poisoning by the soluble mercurial salts and copper. Iron salts are astringent, styptic, and tonic. Although they exert little influence on the skin, the solution applied to mucous or abraded surfaces combine with albumen coagulating the blood, and are powerful astringents. The metal in a finely divided state and the salts, when swallowed, are dissolved by the gastric fluids in the stomach. But, besides restoratives, iron salts are general tonics and astringents; any excess not required for nutrition, reduced in great part to the condition of sulphide, is eliminated in the bile, with the intestinal mucus and the urine, often communicating its dark colour several weeks, the blood and tissues gradually become saturated, and excretion with scarcely any loss occurs in the urine.

Like other irritant poisons, iron acts on the alimentary canal. It induces pain and purging, with coldness of the limbs and surface of the body generally.

Doses.—For horses 1 to 3 drachms, for cattle 2 to 4 drachms, sheep 20 to 30 grains, pigs 10 to 20 grains, dogs 5 to 10 grains. You can either give these doses in the form of a ball mixed with linseed meal and treacle, or you can mix it with soft food, and repeat two or three times a day. This medicine is often given in conjunction with other ingredients. It is valuable for removing worms in horses, when administered along with 2 drachms of aloes. The same ingredients are, along with a quart of ale, a good tonic for cattle. Thirty grains, with 1 drachm each of common salt and nitre, are recommended to be given in sheep-rot. Care should always be taken, during its use, to keep the bowels open by the occasional employment of a laxative, which is especially necessary in the horse and dog, on account of their liability to suffer from the astringent effects and constipating action of the iron salts.

JALAP.

This medicine derives its name from a town in Mexico, where it was first obtained, and from the neighbourhood of which it is still imported. In this country the plant thrives well, flowers, and comes to maturity in the open air.

Jalap is irritant, cathartic, and feebly vermifuge. A full dose given either to dogs or cats causes sickness and vomiting as well as catharsis. When rubbed into the skin it excites inflammation, and when applied to mucous surfaces it is the same. For dogs its action is tolerably certain, and produces full watery discharges, and is especially effective when combined with a grain or two of calomel. It is identical in its effects with scammony; it is more active than senna; but as a purgative for horses it is neither so certain or effective as aloes, and as a purgative for cattle it is superseded by salines and oil.

Doses.—Dogs require 1 to 2 drachms, cats ½ drachm, pigs 1 to 4 drachms. If a dog fasts for six hours, he will be properly purged in two or three hours with ½ to 1 drachm, along with 2 or 3 grains of calomel, made into a pill.

NITRE, SALTPETRE, or NITRATE OF POTASH,

Occurs in the East Indies, Persia, Egypt, Spain, and other warm climates.

A brown incrustation of nitre covers considerable tracts of country. In France and other continental countries, nitre for gunpowder and other purposes is prepared artificially by collecting into large heaps animal and vegetable refuse, with old plaster and other calcareous matter. These heaps are sheltered from rain, but freely exposed to the air; they are frequently watered with urine, and occasionally turned. In about two years the whole is lixiviated and purified by a process similar to that followed with the natural nitre.

Its Action and Uses.—Large doses are irritant and cathartic; medicinal doses are diuretic, alterative, antiseptic, febrifuge, and refrigerant; externally, it is stimulant and refrigerant.

An ounce has proved fatal in the human subject, but very large doses are required to cause serious effects either in horses or cattle.

Nitre possesses a very high diffusive power; it rapidly enters the blood, rendering the venous blood scarlet; it counteracts adhesion of the red globules, favours the solution of fibrine, oxidises products of tissue metamorphosis, and hastens their removal, especially by the kidneys, and neutralises acidity. To one or more of these actions are due its several curative effects. In febrile, inflammatory, and rheumatic complaints it allays fever, lowers excessive temperature, and removes by the kidneys both fluid and solid matters. In the early and acute stages it is combined with other salines and sedatives; in the second stage, with alterative stimulants and tonics. Along with diuretics, it is given in scantiness and turbidity of the urine; and in swelled legs and dropsical affections, small and oft-repeated doses are valuable in arresting purpura in horses, even when iron and turpentine have failed. Most farmers in Scotland give their horses one ounce of nitre in a mash every Saturday night; thus the bowels, kidneys, and skin are kept in good order, and attacks of swelled legs and weeds, so common among hard-worked animals, are prevented or warded off.

Nitre dissolved in water abstracts heat, and is therefore a useful refrigerant.

Doses.—For horses as a diuretic, ½ to 1 ounce, cattle 1 to 2 ounces, sheep 1 to 2 drachms, pigs ½ to 1 drachm, dogs 10 to 30 grains dissolved in water and administered.

CHLORIDE OF SODIUM or COMMON SALT.

Salt is found in Cheshire and Worcestershire, in Poland, Spain, and other European parts. It exists in variable amount in every soil, and therefore in every water. It is the largest saline constituent of the sea, and abounds in the tissues of plants and animals. It is obtained for medicinal and economical purposes by quarrying the solid beds of rock salt, or by evaporating brine springs or sea water. It is soluble in $2\frac{1}{2}$ parts of water, and is rather more than twice as heavy as water.

Action and Uses.—Salt is an essential article of food. Small doses are restorative, stomachic, alterative, and antiseptic; large doses are irritant, cathartic, and emetic. It is used externally as a stimulant, antiseptic, and refrigerant. So essential is the regular and frequent use of salt for the maintenance of health, that animals in a state of nature instinctively travel many miles to saline springs, the sea shore, or beds of salt. The following experiment was tried with half a dozen cattle; these six were selected, and accurate judges were unaminous that their weight was about equal. Well, three of these animals had salt given along with their food, and three had none. In six months' time the three that were deprived of salt presented an extraordinary and striking appearance—the skin and hair became rough and dry and staring; while the other three that received salt presented a decided contrast—their coats were smooth and oily, and although not much superior in weight, they nevertheless brought far better prices when the six were exposed to sale. Those cattle that received the salt exhibited throughout a much better appetite, were more lively; they seemed to relish their food, and consumed it in a shorter time, and also drank large quantities of water. Salt is especially indicated when animals are receiving boiled grains or roots, for the salt naturally present in such prepared food is as a rule in small amount.

During convalescence from acute disease most animals are especially fond of salt. Besides being in itself a restorative, it favours absorption of nutritive matters; therefore animals should have access to salt at all times. A piece of rock salt should always lie in the horse's manger, the ox's crib, and the sheep's trough. The condiment not only gratifies the taste, but in all probability serves other useful purposes. It appears to be the natural stimulant of the digestive organs; it furnishes hydrochloric acid for the gastric juice, and soda salts for the bile; it preserves the fluidity of the blood, and assists in nutrition, for it always abounds where active reparative or formative changes are taking place. It is excreted by the mucous membranes and kidneys.

On horses the laxative action of salt is uncertain, often violent, and usually accompanied by considerable irritation of the kidneys; on dogs it usually operates as an emetic and laxative, and is used to clear out the stomach and intestines, and to induce that soothing action which accompanies the operation of most emetics. Doses insufficient to act on the stomach or bowels are determined to the kidneys, increasing the secretion of urine. On pigs it acts as a purgative, but is scarcely so safe or certain as oil, jalap, and calomel, or aloes.

For vigorous adult cattle and sheep common salt is a decidedly useful and effective purgative; in fact, it is as prompt and powerful as Epsom or Glauber salts. By producing thirst, it compels the animal to drink largely of water, which in torpidity of the bowels, and constipation among cattle, softens and carries onward the hard dry impacted matters, so apt to accumulate in the first stomach and in the manyplies, and resisting the action of ordinary purgatives. Therefore, among cattle and sheep it is administered successfully to evacuate the rumen, to clear the bowels in distention, and in diarrhœa depending on overfeeding, or kept up by the presence of irritating matters in the canal. It is given to relieve irritation and inflammation of the eyes, brain, respiratory organs, and limbs, and in such cases bowels, freeing the blood from peccant matter, and exciting counter irritation. It is the best antidote for silver salts. Small and repeated doses are stomachic, and are prescribed with gentian, ginger, or spirits and water, for all animals suffering from indigestion. Salt given regularly lessens the liability to intestinal worms, and an injection of half an ounce in a pint of warm water will destroy those ascarides that lodge in the rectum. It obviates in a great measure the evil effects of damp and badly kept fodder, and prevents or retards the progress of liver-rot in sheep. From its action as a stimulant, as well as from the cold it produces during solution, it is of great benefit in many diseases of the feet and joints, particularly among cattle and sheep. When a freezing mixture is required, one part of salt is mixed with two parts of pounded ice. However these must be used with great caution, for if applied for many minutes at a time, they are apt to lower the vitality to a dangerous extent.

Medicinal Doses.—For a purge the ox and cow take from ¾ to 1 lb., sheep 1 to 3 ounces. It is, however, better given along with Epsom salt in half the quantity each, adding 2 ounces of powdered ginger and 1 lb. of treacle. Should the dose fail to act in twelve hours, it can be repeated, adding a pint of linseed oil. As an alterative for horses and cattle, 1 or 2 ounces are given; as an emetic for the dog, 1 to 4 drachms. But a still more effective vomit for the dog is a tablespoonful of salt, and a teaspoonful of mustard dissolved in 3 or 4 ounces of water.

SUGAR

Is present in many plants. It is manufactured in France from beetroot, and in America from sugar maple. But the sugar employed in this country is obtained from the sugar cane which is cultivated largely in the West Indies. The lower parts of the cane are richest in saccharine matter. These canes are crushed between heavy rollers, and the juice is thus expressed, which contains nearly 20 per cent. of sugar; it is mixed with a little slaked lime to neutralise acids and precipitate albuminoids, and concentrated in shallow vacuum pans, at a temperature not exceeding 140°; the coagulating albumen which entangles impurities is then skimmed off, the syrup is cooled down in wooden vats, and dried in the sun, when yellow dark brown raw sugar is formed: 1 cwt. of raw sugar gives about 80 lbs. of refined sugar and about 16 lbs. of treacle.

Its Actions and Uses.—Sugar is readily absorbed, and acts as a respiratory fuel, or it is converted into and stored away in the system as fat: 1 or 2 lbs. given to a horse, 8 to 12 ounces to dogs, increase the amount of fluidity of the fæces and augment the secretion of the urine. Treacle is often substituted for sugar. It is palatable, digestible, laxative, and well adapted for sick animals. It is a convenient purgative, and covers the disagreeable flavour of salts and other laxatives. When full doses of physic have been given, and their repetition is dangerous, repeated doses of treacle can be had recourse to with safety, especially in cattle and sheep. As a gargle for horses with sore throat, 4 ounces of treacle and 1 of nitre, dissolved in a pint of water, and slowly given every hour or two, is of inestimable value; it is simple, and is found in almost every home where stock are kept.

Doses of Sugar and Treacle as a Purge.—Horses and cattle require 1 to 2 lbs., sheep 3 to 4 ounces, pigs 2 to 3 ounces, dogs 1 to 2 drachms, given in water, ale, or gruel.

SULPHUR

Is one of the old-established remedies that used to cure everything in former times. It is known and recognised by its characteristic pale yellow colour. When heated it easily inflames, and in burning evolves the highly irritating fumes of sulphurous acid, thereby asserting its disinfectant properties. In cases of fevers of an infectious nature, the authorities in London close up the houses, exclude all air from entering either by doors or windows, and then set fire to a large quantity of brimstone or sulphur, and allow it to burn for twenty-four hours; at the expiration of that time, the occupant receives the key of his house. During the interval they care not what becomes of the tenant; they have this duty to perform, and they carry it out most religiously to the very letter of the Act. I am speaking from experience upon the subject, although it is a painful reminder.

The Action and Uses of Sulphur.—In large doses it is an irritant poison; in medicinal quantities it is a laxative, alterative, and stimulates the mucous surfaces; applied externally, it is an efficient antiparasitic and stimulant to the skin, and is fatal to insect life.

It is administered to several domestic animals, to open the bowels gently in pregnancy or in piles, where a more powerful and active purgative would cause irritation. It is also given in chronic chest disorders, in convalescence from acute diseases, and sometimes in rheumatism and skin affections.

Doses as a Purge.—Horse require from 3 to 4 ounces, cattle 4 to 6, sheep and pigs 1 to 2 ounces, dogs 2 to 4 drachms.

SWEET SPIRIT OF NITRE.

For all animals this is indeed a valuable and serviceable
stimulant ; it is a ready rouser of the heart's action, an
inestimable carminative and antispasmodic in indigestion,
tympanitis, and colic ; and further, it is of inherent importance in typhoid fever, and in convalescence from debilitating
diseases: It is an effectual excitant to the skin and
kidneys in cold, rheumatism, and local congestion. Like
alcohol and ether, properly regulated doses lower the excessive animal temperature, and antagonise blood-poisoning.
It is principally thrown off by the lungs.

Large doses are narcotic, producing delirium and coma,
with a variable amount of preliminary excitement. In
medicinal doses it is a stimulant, antispasmodic, diaphoretic,
diuretic, and antiseptic.

Doses as a Stimulant and Antispasmodic.—Give to a
horse 1 to 2 ounces, to cattle 1 to 4 ounces, to sheep 2
to 4 drachms, to pigs 1 to 2 drachms, and to dogs 15 drops
to 1 drachm. It should never be mixed with anything
until it is to be administered. It is usually given in water,
ale, or linseed tea. A horse suffering from colic requires
2 ounces with 1 ounce of laudanum, and with 5 or 6
drachms of aloes in Scotland added, but in England 2 or 3
drachms of aloes will suffice. It is difficult to reconcile the
diversity of these doses in the north and south; nevertheless,
such is the case. We can give an 8 and even a 9 drachm
ball to a Scotch horse, when the same dose would kill an
English one. Sweet spirit of nitre 2 ounces and the same
quantity of laudanum is of especial value to cows after
calving, as it counteracts and arrests the spasms which
usually follow the birth. For influenza and typhoid fever
in horses, 2 ounces each of sweet spirit of nitre, acetate of

ammonia, and a drachm of belladonna extract, form a splendid anodyne draught, given in gruel or water.

SULPHATE OF MAGNESIA or EPSOM SALT.

This highly important salt is found in various rocks and soils in the ocean, in the proportion of 15 to 20 grains in a pint, and in some mineral springs. Owing to its presence in the streams at Epsom, it derives its name.

It possesses a cooling, saline, nauseous, bitter taste; is insoluble in alcohol, but soluble in its own weight of temperate water. It is a purgative, alterative, febrifuge, and antiphlogistic; it closely resembles common and Glauber salts, and is a more active purgative than potassium bitartrate or phosphate of sodium. It has a low diffusive power, passes tardily through animal membranes, retards absorption of fluid from the canal, and is itself chiefly absorbed from the stomach and small intestines.

Whilst in the blood it is believed to diminish the cohesive tendency of the red corpuscles, to remove fibrin, and retard the coagulation; it is shortly excreted through the glands of the bowels and vessels into the large intestines, along with a considerable amount of serum and fibrin in solution. It counteracts biliousness alike in man and animals, by sweeping away the unabsorbed bile generally present in the intestines, and which, unless removed, becomes reabsorbed. Like other salines it is uncertain, and at other times too violent for horses; it often acts unexpectedly on the kidneys, but in repeated doses of 2 or 3 ounces, it proves a valuable alterative and febrifuge. On dogs its purgative effects are irregular, and frequently accompanied with nausea and vomiting; but for cattle and sheep it is a most excellent and convenient purge, equalled only by common salt in rapidity and fulness of action. When administered to cattle full doses generally take effect in twelve or fifteen hours, and cause very fluid evacuations. Besides producing purgation, it diminishes blood pressure, and abates fever and plethora; while moderate and repeated doses in all animals also augment the secretions of the skin and kidneys.

Medicinally, it is given to all animals that chew the cud for the ordinary purposes of a purgative, to evacuate the bowels in indigestion, constipation, and diarrhœa; to remove noxious matters from the blood, as in febrile and inflammatory affections; and to induce extensive counter irritation, as in inflammation of the brain, eye, and most other organs, the intestines excepted. Although not a desirable purge for horses, it is a good febrifuge: 1 to 3 ounces given in pneumonia influenza, and in most inflammatory disorders improve the appetite, remove the clamminess of the mouth, lessen fever, lower blood pressure and excessive temperature, and help to establish and maintain a healthy and regular action of the bowels. For such purposes it is given twice a day, whether for horses or cattle, but it must be withheld or diminished in quantity when the bowels become relaxed, or where flatulence or spasm follows its administration. It is often combined with nitre and other salines, and given in convalescence from acute diseases, with powdered gentian and other carminatives. Epsom salt is one of the best antidotes for lead poisoning; it converts it into an insoluble sulphate, and further evokes the action of the bowels, which in lead poisoning is liable to be impaired and tardy.

Doses as a Purge.—Cattle require 1 to 2 lbs., calves three months old 3 to 4 ounces, sheep and pigs 4 to 6 ounces, dogs 2 to 4 drachms, dissolved in ten or fifteen parts of water. It is always better to add treacle to it, as this conceals its bitter taste. In obstinate constipation in cattle, it is often necessary to add croton beans, say 12 or 15.

ALOES.

This is the most valuable purgative that can be administered to the horse; and although there are numerous varieties of this medicine, I shall only speak of the true Barbadoes, as it is the only one which you can place implicit reliance upon, and the one most extensively in use.

Barbadoes aloes has a dark or liver brown colour; a brown, opaque, earthy fracture; a disagreeable, bitter, persistent, and a strong and unpleasant odour, especially when you breathe upon it. It is tough, hard, and very difficult to powder; small fragments are translucent, and of an orange-brown hue; its powder is olive green, and darker than that of the other commercial varieties. The dark colour, dulness, and opacity of Barbadoes aloes are generally stated to depend upon the presence of water, but may also be owing to the condition of the aloin. When dissolved in weak spirits, it leaves an abundant flocculent residue.

Its Action and Uses.—Considerable doses are purgative; repeated small doses, insufficient to increase the action of the bowels, are tonic; applied externally, it is stimulant and desiccant. Aloes, when given in the solid form by the mouth is emulsified and saponified chiefly by the bile and pancreatic fluids, and then in great part absorbed. The rapidity with which a properly compounded ball dissolves in the horse's stomach is astonishing; experiments have proved that a ball will be entirely dissolved within an hour after administration. Aloes has not as yet been found in the blood, but it has been found in the milk and other secretions; and its frequent action on the kidneys is almost certain evidence that it enters the circulation also. Being, however, a foreign body, it is quickly excreted or thrown off. Insoluble in air, it is not removed by the skin or lungs. In full medicinal doses, it is not easily separable by the kidneys, but is specially attracted to the glandular apparatus covering the intestinal mucous membrane; it induces there irritation, and a copious secretion which is poured into the canal. Compared with some of the other purgatives, aloes is tardy in its action, and liable to be uncertain when the bowels are irregular or loaded with hard, indigestible dry food; but it is a safe, certain, and sure purgative for horses. Unlike many other cathartics, and excepting in inflammation of the alimentary mucous membrane, it is not in large amount an irritant poison unless it is given in very large doses; it does not render the evacuations so fluid as saline purges, but it appears to increase in degree the movements of the bowels.

In the horse, a purgative dose of aloes produces in a few hours dryness and increased warmth of the mouth, the temperature is increased, the pulse is quickened, and occasionally sickness with colic, and profuse staling. This diuretic effect occurs with the best Barbadoes aloes, more especially if the bowels have been in a state of constipation previously, or otherwise out of order. When the drug is combined with ginger or other aromatics it helps to ward off and prevent these effects. Now the time required for the operation of aloes differs materially in different horses, and is modified by various circumstances, more especially by the sort of diet that the animal has been previously kept on, therefore in some cases the medicine will produce purgation in sixteen hours, and in others it will require twenty-four or twenty-six. Again, in some horses the purging is over in two or three hours, and in others it will extend to twenty-four.

In cattle it is neither prompt certain, nor powerful, even when given in the fluid state, and in doses of several ounces. Six ounces have ere now been given to a cow without producing any effect—in fact, the bowels remained unmoved; and as a purge for the dog it is neither safe nor speedy. It is, however, a tolerably good purge for swine, although it takes from twelve to fifteen hours to operate.

It is administered to the horse in colic, constipation, and indigestion, and for the expulsion of worms and foreign substances from the intestines. For colic, the late Professor Dick of Edinburgh recommended 5 drachms of aloes dissolved in a quart of hot water, and given with an ounce each of oil of turpentine and laudanum.

It is a valuable remedy in treating inflammation of the brain, eye, absorbents, and joints, and owes its efficacy to one or other of the following causes. It clears the stomach and intestines of indigestible food, it removes from the blood noxious matters, it establishes counter irritation, and by reflex action promotes a healthier state of diseased parts. It is also effectual in removing enlargements and dropsies when they do not depend on debility or disease of important internal organs. Repeated doses will reduce superfluous blood and fat, but this object can be more judiciously accomplished by careful feeding and well regulated exercise.

Doses.—Horses can take from 2 to 9 drachms, or even 10 in the north; cattle take from 1 to 2 ounces, sheep ½ to 1 ounce, dogs 30 grains to 1½ drachms, swine 2 to 5 drachms.

ACETIC ACID.

Concentrated acetic acid is an active irritant poison in any animal. A drachm is sufficient to kill a rabbit in four hours, and 1 ounce in seven minutes: 6 drachms of vinegar killed a rabbit in eight hours. In the animals experimented on by different observers, the effects of acetic acid or vinegar in poisonous doses have been great abdominal pain, vomiting, prostration, feeble respiration, and a small frequent pulse. In the cow the following symptoms were developed— a disposition to lie, great difficulty in standing, hesitating gait, laboured and sonorous breathing, small and frequent pulse (136 per minute), temperature of the body variable, soft excrement, and milk arrested, the appetite entirely lost, abdominal pains very severe, great listlessness, convulsions, and death.

Post-mortem Appearances.—On opening the stomachs and intestines, the odour of vinegar indicates the nature of the irritant that has produced the foregoing symptoms during life. The mucous membrane is soft, and of a dark red or coffee colour. Strong acetic acid induces solution of the tissues and hæmorrhage.

Rubbed into the skin, acetic acid speedily causes redness, and the eruption of large blisters, resembling those produced by boiling water; but, as a blistering agent, mustard or Spanish fly is far more desirable. Vinegar, along with either hot or cold water is a convenient stimulant for superficial inflammation, sprains, and bruises, and a refreshing antiseptic for sponging the body in fever. It removes warts as well as corns in the human subject, it softens scurf, destroys parasites and acari, and is successfully employed in ringworm, scab, and mange.

Given as a medicine, the dose is 1 to 2 drachms of acetic acid diluted well with water for horses and cattle, 10 to 20 drops for sheep and pigs, and 2 to 5 drops for the dog.

ALUM

Is slightly irritant and astringent, and is principally employed externally as a styptic, desiccant, and astringent: 1 or 2 ounces given to dogs causes vomiting, and when the gullet is tied, it produces death.

Post-mortem Appearances.—Internal inflammation of the mucous membrane of the intestines.

Treatment.—Calcined magnesia in water.

NUX VOMICA—*STRYCHNOS.*

A plant of the order *Loganiaceæ*, the seed of which is the deadly *Nux-vomica.* The tree is of moderate size, and grows in Ceylon and several districts in India. It has a short crooked stem, ribbed leaves, small greenish white flowers, and a beautiful round orange-coloured fruit the size of a small apple, having a brittle shell, and a white gelatinous pulp. The wood is intensely bitter, particularly that of the root, which is used to cure intermittent fevers and the bites of venomous snakes. The seeds are employed in the distillation of country spirits, to render them more intoxicating. The pulp of the fruit seems to be perfectly innocent, as it is greatly eaten by many sorts of birds. The seeds are circular, not quite an inch in diameter and two lines in thickness, concave on one side and convex on the other; they are very tough and horny, covered with a velvety down consisting of fine hairs, ash-coloured, and silky. Internally, the seeds are whitish and translucid; they are very difficult to reduce to powder, possessing no odour, but are extremely bitter. The Germans fancy they can discern a resemblance in them to grey eyes, therefore they

D

call them crows' eyes. Dog killer and fish scale are two Arabic names for the vomic nut.

For practical purposes, strychnia may be regarded as the active principle of *Nux-vomica*.

Whether the *Nux-vomica*, or its alkaloid strychnia, is used, the effects are the same, the only difference is in the dose. A horse has taken from 1 to 3 ounces of *Nux-vomica* with impunity. But Vallon, on the other hand, asserts that from 6 to $7\frac{1}{2}$ drachms invariably prove destructive; while 10 grains of strychnia are sometimes more than sufficient to kill a horse. From 3 to 4 grains introduced into the cellular tissue, and any dose above half a grain injected into a vein, will prove fatal. Cattle will support much larger doses of the *Nux-vomica* than the horse. According to authority, 4 grains of strychnia introduced beneath the cellular tissue destroyed a cow in twenty minutes; and with regard to the smaller ruminants, Hertwig says that a goat two years old partook of upwards of four ounces of *Nux-vomica* in the course of eleven days, without manifesting any signs of suffering. Dr Christison of Edinburgh has seen a wild boar killed in ten minutes with the third of a grain, injected in the form of an alkaline solution into the chest. Dogs, if very robust and large, can be destroyed with half an ounce of *Nux-vomica*, and of strychnia half a grain blown into the mouth of a dog produced death in five minutes. Again, Dr Christison says he has killed a dog with the sixth part of a grain, injected in the form of an alkaline solution into the chest.

The *Symptoms* are as follow:—A few minutes after the introduction of the poison the animal becomes agitated, and trembles; in a short time it is seized with stiffness and starting of the limbs, which increase until a violent general spasm ensues, in which the head is bent back, the limbs are extended and rigid, the spine stiffened, and respiration checked, the chest being fixed. The slightest noise and touching the animal exites fits; during the latter, there is occasionally involuntary emission of urine. Intervals of rest occur, but the mucous membranes acquire a red colour, the pulse is quick and hard, paroxysm follows paroxysm until the animal perishes, suffocated or exhausted totally.

The Post-mortem Appearances.—The left auricle of the heart and also the intestines have been known to contract for an hour after death. The lesions are those met with in cases of death by suffocation. The viscera have been found perfectly healthy. The brain and spinal chord injected, and fluid accumulated in the spinal canal.

Treatment when the Poison is in the Stomach.—Vomiting must be excited in those amimals that can thus evacuate the stomach. Oleaginous draughts and purgatives must be given; alcohol, ammonia, sulphuric ether, and camphor have all proved useful. Morphia and opiates likewise act beneficially, and chlorinated water has been recommended as an infallible remedy. Artificial respiration ought to be persevered with, and infusions of gall and green tea, on account of the tannin they contain, are said to be valuable antidotes.

The Tests for Strychnine.—Dr Stevenson Macadam, Professor of Chemistry, Edinburgh, has made some very interesting experiments on strychnine poisoning and the tests for strychnia. The tests are many, and some quite characteristic, as will be noticed by the following table :—

" *The Strychnine Tests.*

A. Potass, a white precipitate, insoluble in excess.

B. Bicarbonate of soda (in acid solution), no precipitate.

C. Sulphocyanide of potassium, a white precipitate.

D. Perchloride of mercury, a white precipitate.

E. Perchloride of gold, a lemon yellow precipitate.

F. Chlorine water, a white precipitate, which dissolves in ammonia to a colourless liquid.

G. Nitric acid (cold), colourless solution (heat), a yellow solution.

H. Sulphuric acid (with trace of nitric acid), and binoxide of lead, a violet changing to a red colour.

I. Sulphuric acid and binoxide of manganese, a violet changing to a red colour.

J. Sulphuric acid and bichromate of potash, a violet changing to a red colour.

"The tests *A* to *G* cannot be applied, excepting when the quantity of strychnine at the command of the operator is considerable, so that in dilute solutions they fail to act. The remaining tests, *H* to *I*, are however much more delicate, and will indicate a most minute amount of strychnine."

To detect the substance, digest a part of the stomach or other substances supposed to contain strychnine in a dilute solution of oxalic acid for some hours, thereafter warm and strain through muslin. The filtrate is rendered slightly alkaline by stirring with a rod of caustic potash. It is then placed in a stoppered narrow-necked bottle, several ounces of ether are added, and the whole is well shaken. The liquid is allowed to settle, when the ether will rise to the surface with strychnine in solution (that is to say if any be present). The ether is then drawn off into a porcelain evaporating basin, or even in a common porcelain plate ; it is allowed to evaporate spontaneously. When nearly dry, heat is applied to remove any traces of remaining ether, and the residue is tested for strychnine in the following manner : —A little of the residuum is tasted; if the taste be strong and bitter, strychnine is very likely present. A few drops of the strongest sulphuric acid are placed upon the plate, and a drop of solution of bichromate of potash is added; the two substances are then allowed to run together, when, if strychnine be present, beautiful violet streaks will be perceived, which soon change to red. Therefore this test is quite sufficient to identify strychnine.

Dr Macadam says—"So far as my experience goes, I prefer the sulphuric acid and bichromate of potash test, as it is much more certain in its action and is more delicate than any or all of the other tests. The colour indications are best seen in a pure solution of strychnine; the presence of organic matter impedes the action of the test, and alcohol, acetic acid, and other bodies, entirely destroy the characteristic colour. In order to steer clear of these sources of error, Dr Letheby has lately suggested that the substance to be tested should be treated with sulphuric acid, and placed on a piece of platinum foil connected with the positive pole of a galvanic battery, and thereafter, on touching the liquid with the negative pole of the battery, which terminates in a platinum wire, the characteristic violet tint is at once perceived and produced. In this way $\frac{1}{10000}$th of strychnine in pure water has been detected. I have repeatedly tried this process, and can bear witness to the accuracy of the test; but in practice I have found the sulphuric acid and bichromate of potash to be a more delicate test, though it is much more difficult to manage. Lately a good deal has been said in disparagement of the colour tests for strychnine, and considerable doubt has been thrown upon the trustworthiness of colour tests in general. Precipitate tests are certainly more satisfactory than colour tests, because they signify the presence of a larger amount of the particular substance under examination, but in general colour tests are far more delicate in their action than precipitate tests."

" A very good example of this occurs in testing for iodides. When these are abundant, precipitate tests with soluble salts of lead and mercury may be readily obtained; but by dilution, a point is at last reached, when lead or mercury solutions cease to be precipitated by the liquid containing the iodide. At this point, the starch test comes into play, which in a very dilute solution of an iodide is essentially a colour test; and long after the precipitate tests fail to indicate an iodide, the colour test shows unmistakable evidence of its presence. The same remark applies to testing for solutions of persalts of iron and copper, by means of ferrocyanide of potassium. In strong solutions, a blue precipitate is indicative of iron, and a ruddy brown precipitate speaks of copper; but when dilute solutions are examined, blue and ruddy brown colorisations are alone obtained. Colour tests, therefore, are the most delicate of all tests; they indicate the presence of a body, when precipitate tests cannot do so, and, for my own part, I see no reason why I should distrust my sense of colour, whilst manipulating in my laboratory, and confide in it at other times."

Doses of Nux-vomica.—Horses take 1 drachm, cattle 2 to 3 drachms, sheep 20 to 40 grains, pigs 10 to 20 grains, dogs 2 to 8 grains, repeated twice daily for a week or ten days.

Doses of Strychnine.—It must be remembered that this agent is about ten times more active than *Nux-vomica,* and more than thirty times as active as the powdered nux. The horse takes 2 to 3 grains, cattle 4 to 6 grains, sheep one-third of a grain to 1 grain, and for dogs one-third to one-tenth part of a grain. I have administered it in paralysis in the cow with the best possible results, in dogs after distemper, and in all animals after lead-poisoning.

TARTRATE OF ANTIMONY—TARTAR EMETIC.

Administered to dogs and other carnivora, it is an irritant poison. It is still given by coachmen and grooms as an alterative for horses, and many fatal results have arisen from its use in this way, although large doses can be sustained by the horse. Hertwig asserts that 2 ounces are sufficient to destroy life. Experiments instituted at Alfort show that 4 ounces produced death only on the third day. Mr Barlow, along with Mr Dun, performed some very interesting experiments with this poison. "A brown mare died from the effects of 86 drachms of tartar emetic, taken in six days. Another mare, 16 hands high, took 83 drachms in eighteen days, but without exhibiting any physiological effect. A black mare of sound healthy constitution, took 84 drachms in doses of 4 drachms, repeated twice a day, during the ten days from the 16th to the 26th September 1852, and she improved in condition, and was in no way affected by it. A healthy well bred horse received 10 ounces of tartar emetic in solution, and after showing a good deal of nausea, uneasiness, and pain, died in about six hours. The only notable appearances on *post-mortem* examination were softness and vascularity of the intestines, analogous to what is seen in patients that have died while affected by diarrhœa. Neither in this nor in any other of the cases were the lungs congested or inflamed, as is said to have occurred in Magendie's experiments."

Mr Balfour of Kirkcaldy says, that he has given half a pound in solution to cattle without any very obvious effects. Hertwig says, a quarter of an ounce was sufficient to destroy an old pig, and 40 grains given in two days killed a pig five months old.

If the dog's gullet be tied, from 4 to 6 grains will kill;

but if he is allowed to vomit, 8 scruples can be given, and death not follow.

Symptoms of Poisoning.—Vomiting, diarrhœa, vertigo, thirst, salivation, dulness, depression, cold and clammy skin, colic, spasmodic contraction of the muscles, convulsions, sometimes paralysis of the hind quarters, and death.

Post-mortem Appearances.—General inflammation of the alimentary canal, sometimes ulceration. If the tartar emetic has been given in a solid state, occasionally an eruption on the mucous membrane is observed, similar to that of variola ovina or small-pox, lungs congested, blood dark and fluid, ecchymosis in the heart, &c.

Treatment.—Vegetable astringents combine with the oxide of antimony to form insoluble compounds, so that tannin, galls, oak bark, Peruvian bark, catechu, and even strong tea, may prove efficacious ; if vomiting and purging continue, an injection containing opium should be given.

Doses, as an alterative and sedative, for horses or cattle 1 to 4 drachms three times a day, either in a ball or solution. As an emetic, for dogs or cats 1 to 4 grains, and for pigs 4 to 10 grains.

ARSENIC.

Most people are now cognizant of the deadly effects of arsenic, which makes one the more surprised that extra precautions should be neglected when occasion requires its employment. Only a short time ago, a farmer in the neighbourhood of Lauder, N.B., lost several valuable cows by the external application of arsenic and sulphur to the animals, which were affected with lice. Nor is this an isolated instance, for numerous cases here and there are ever cropping up, all of which testify that due regard to its poisonous character is either imperfectly known or contemptibly ignored.

Arsenic acts on all animals as a destructive, irritant, corrosive poison. It produces irritation, inflammation, and sloughing of any mucous or abraded skin surface with which it comes in contact; it is quickly absorbed, and produces, while it remains in the system, loss of appetite, emaciation, various nervous disorders, and depression of the circulation; it is evacuated by the stomach, intestines, also by the liver, and to a large extent by the kidneys, causing extensive irritation to these channels as it passes through them. It exerts its poisonous influence with equal certainty by whatever medium it enters the body. All its compounds are poisonous, and as usual with other poisons, the most soluble are the most deadly. It has been proved that doses from 40 to 70 grains destroyed dogs in from two to six days, with much the same effects, whether they were swallowed or applied to a wound. Metallic arsenic, although itself innocuous, unites readily with hydrogen, oxygen, and other bases, speedily acquiring poisonous activity. White arsenic, like other poisons, has often been administered with impunity in considerable doses to horses. An animal suffering from inveterate mange has had 3 drachms without any perceptible injury. A horse affected with glanders began with 1 drachm daily, made into a ball with linseed meal and treacle; this dose was increased by 20 grains per day; at this period the animal got in one dose 380 grains, and then had taken upwards of half a pound of arsenic. Yet, notwithstanding this extraordinary quantity, no effect was obvious; there was no uneasiness, no pain, and no alteration of the pulse or breathing.

But, although such large doses have sometimes little effect, much smaller ones occasionally exert greater violence. For example, 5 grains administered daily to a horse in the form of a ball produced shivering, loss of appetite, purging, and other symptoms of abdominal irritation, imperceptible pulse, prostration of strength, and finally death on the ninth day.

These very different effects, however, depend greatly upon the varying susceptibility and on the amount of food present in the stomach and intestines, and on the fact that animals acquire a tolerance of it when they receive it regularly. To substantiate these remarks, it must be remembered that arsenic administered in solution (as it ought to be), when given internally, is more certain, active, and regular than when supplied in the solid state.

Dipping mixtures which contain arsenic often produce serious and fatal consequences by inattention to instructions, neglect, or carelessness. In 1858 a case occurred in Northumberland during the summer which excited intense interest amongst chemists, agriculturists, and veterinary surgeons. The particulars are as follows:—Mr Elliot, chemist, Berwick-on-Tweed, sold to Mr Black of Burton fifteen packets of sheep-dipping mixture : one of these packets contained 20 ounces each of arsenic and soda ash, with 2 ounces of sulphur. This was directed to be dissolved with 4 lbs. of soft soap in 3 or 4 gallons of boiling water, adding after this 45 gallons of cold water, which made enough of the dip for fifty sheep. About the middle of August Mr Black dipped 869 sheep in the usual manner. The whole appliances and arrangements were good, while every care was taken with the dipping. In two days, however, the sheep began to die, and within a month 850 had succumbed. The usual symptoms of poisoning by arsenic were exhibited, namely, frothing at the mouth, blood-shot eyes, pain in the bowels, black and bloody urine, laboured breathing, blackening of the skin, wool falling off in patches; and on analysis of the stomachs and bowels, &c., arsenic was found.

Mr Elliot, on the other hand, proved that thousands of sheep had been dipped in mixtures of the same strength as that supplied to Mr Black with impunity; and in support of this statement, another gentleman in the vicinity employed eight packages of the very same mixture, made in the same way, and at the same time; and a professional sheep-dipper from the south of England, who annually passes through his hands thousands of sheep without losing one, has been in the habit of employing 2½ lbs. of arsenic for every fifty sheep, exactly double the strength of Mr Elliot's mixture. Yet with such evidence the jury found a verdict for Mr Black, and assessed the damages at £1400.

Now the conclusions to be drawn from the aforementioned case is, that arsenical sheep-dipping mixtures are not liable to be absorbed through a healthy skin. The risk in using such dips depends, not on their being taken into the system by absorption, but on a quantity of the poisonous fluid being retained in the fleece, so that the first shower of rain washes the fluid from the wool on to the grass or other food over which the animal strays. This undoubtedly explains the great mortality which occurred at Mr Black's. Therefore due precaution is absolutely necessary when arsenical dressings are employed either as ointments or lotions.

Doses.—Horses and cattle take from 5 to 10 grains, sheep 1 to 2 grains, dogs one-fifteenth to one-tenth of a grain.

SULPHATE OF COPPER or BLUE STONE.

All the salts of copper are poisonous. They may destroy life if used as caustics, or when large doses are introduced into the stomach. Medicinally, however, it is an astringent, tonic, and antiseptic ; and for carnivora it is a prompt emetic. Externally, it is used as a stimulant, astringent, and caustic; while it is an effectual antiseptic.

As a tonic, horses take from 1 to 2 drachms, cattle 1 to 4 drachms, sheep 20 to 30 grains, pigs 5 to 10 grains, dogs ¼ to 2 grains. These doses can be repeated twice a day. It is a useful ointment in foot-rot in sheep. Take equal weights of powdered blue stone, gunpowder, and lard, mix and apply; or the following is a more adhesive application, mix over a slow fire one part of powdered blue stone with three parts of tar, and apply to the feet when cool.

CREASOTE.

A mixture of volatile oils obtained by the distillation of wood tar, B.P. It is present in wood and peat smoke. Tar of good quality contains from 20 to 25 per cent. of creasote. It has been employed of late years in the treatment of the diseases affecting cattle. It is an active caustic or corrosive, and, in virtue of these properties, has been strongly recommended for canker in the foot. One to 2 drachms of creasote given internally to a dog induces great anxiety, staring look, debility, and even paralysis of the extremities, vertigo, vomiting of a white coagulated substance, bloody foam at the mouth, loud breathing, and symptoms of suffocation, ending in death. On opening the body, a strong odour of creasote, like that of smoked meat, is detected in all viscera, the mucous membranes of the stomach and intestines are of a dull red colour and inflamed, and in some parts corroded, while the blood is thick and black. The same symptoms have been observed to supervene, the above mentioned dose of creasote being mixed with an equal weight of water.

Its Medicinal Action.—Small repeated doses arrest undue fermentation in the stomach, and hence it is useful in some forms of indigestion, and in dogs in chronic vomiting.

A few drops inhaled with steam proves soothing in bronchitis and chronic lung complaints, especially when accompanied with excessive fœtid discharges.

At Dick's Royal Veterinary College, Edinburgh, some years ago, creasote was used in many cases of contagious. pleuro-pneumonia among cattle, and in doses varying from 20 to 80 drops, dissolved in volatile oil or acetic acid, and with some temporary advantage in relieving the distressed breathing and irritable bowels. It has also been tried in glanders, and like all other articles has been found wanting. As an antiseptic it stands next in order to carbolic acid, and it is believed to have been the essential agent used in embalming the Egyptian mummies.

Doses.—Horses take 20 to 40 drops, cattle ½ to 2 drachms, sheep 10 to 20 drops, pigs 5 to 10, and dogs 1 to 3 drops.

It is given with syrup or alcohol, &c.

MUSTY HAY.

When a wet season interferes with the proper making and drying of hay, it is a sure precursor of many diseases of the digestive organs, and especially of stomach staggers and colic. In Scotland, about the end of 1856, there was. great mortality amongst horses from stomach staggers, induced by bad hay.

Symptoms.—They move along with hanging head, sunken eye, depressed lip, and tottering gait, suffering from pain in the bowels, with considerable swelling, partial sweats bedew the body, the visible mucous membranes are highly injected, and the urine of dark colour. The muscles twitch, the animal writhes in pain, and dashes itself about in fits of delirium. In Scotland, when the great outbreak occurred, the disease was followed by partial paralysis of the hind legs.

Treatment.—There is no use lingering about the bush; give at once an active dose of physic, for upon this course of action does the life of the patient depend. It is all humbug to think that bleeding will effect a cure in this complaint. It never has done so yet, and never will. Because you have a large quantity of indigestible material lodging in the stomach and bowels, which is the direct cause of the ailment, therefore, to be successful, remove the cause and the effect will cease; the cause is the musty hay. A good dose of physic then clears out this effete material, and when that is accomplished your horse is on the royal road to recovery. Bad straw has also the same pernicious effects, because one cannot help a horse from eating it at night. However, these diseases are preventable; so prevention being better than cure, such feeding stuffs ought to be avoided. Should this ensilage process prove effectual, it will be a valuable boon to the owners of stock in more ways than one.

LINSEED.

Linseed Oil.—The oil expressed without heat from linseed.

Linseed Meal.—The cake of linseed from which the oil has been pressed and reduced to powder, B. P.

Linseed and linseed cakes are extensively used and highly valued, owing to their nutritive qualities as feeding stuffs. They are capital fat-producers, are palatable and digestible, and are more effective than either sugar or starch. When well boiled it is a most excellent food for horses, cattle, and sheep, not only in health, but in disease also. It is both a nutritive food and medicine. A daily mash of linseed gruel, or a small quantity of cake given to lambs and calves, as soon as they can eat it, not only favours their growth and development, but is successful in preventing attacks of diarrhœa, dysentery, black leg or black quarter. In the shape of gruel, it is simply invaluable in irritable conditions of the throat, bowels, kidneys, and bladder, also in poisoning by irritants and corrosives. It is a convenient vehicle for the administration of nauseous or acrid medicines. Ground linseed makes the best poultices; it is in common use for making up balls with treacle. Linseed oil has been deservedly praised, but it does not answer very well for cattle or sheep. The cake possesses the advantage as a feeding stuff. As a restorative, however, in the horse, 2 ounces daily of the oil is often beneficial in sore throat, bronchitis, &c. In fact, I know plenty of farmers who are in the habit of giving a wine glassful every night in the horses' feed, and so accustomed have those animals become to it that they absolutely refuse their food if the oil is awanting.

One of the colic draughts of the Edinburgh Veterinary College is, linseed oil 1 pint, with 1 or 2 ounces each laudanum, and oil of turpentine. It often relieves choking in cattle, and as an injection it allays irritation in the rectum. As a soothing dressing, it is applied to hard, dry irritable surfaces.

Doses as a Purgative.—Horses take 1 pint, cattle 1 to 2 pints, sheep and pigs 6 to 8 ounces, dogs 1 to 2 ounces, cats 1 drachm, given in milk or spirits and water, and well shaken up before administration.

CASTOR OIL.

The oil expressed from the seed of *Ricinus communis,* imported chiefly from Calcutta, B.P. This plant is generally considered to be Jonah's gourd. It is cultivated in the colder parts of Europe. It is an annual shrub from 4 to 5 feet high ; but in Spain and Sicily it reaches the height of 20 feet, and in North and South America it becomes a large tree.

Castor oil seeds are irritant and purgative, and have frequently caused fatal inflammation of the stomach and bowels in man. They appear to be more powerfully irritant than the oil extracted from them ; when crushed they form an Indian cure for mange.

The oil is a mild purgative, closely resembling linseed and the other fixed oils. It very rarely causes griping, and increases the motions as well as the secretions of the intestines. When injected into the veins of man, it causes laxative effects, and produces in the mouth the same disagreeable oily taste as when swallowed. In the horse, castor oil is not very certain or prompt in its effects, and is seldom to be recommended where a speedy and full purgative effect is desired. In cattle and sheep its action is more reliable ; but in these animals it is superseded by linseed oil, which is similar in its action, and a good deal cheaper. In the dog, however, it is more active than in man; and for delicate subjects, a mixture of equal quantities of castor oil and olive oil is preferable to castor oil alone. Its nauseous oily taste is obviated by beating it up with an egg, or a little spirits, or some aromatic. Acting as it does without irritation or griping, it is useful in young animals in irritation and inflammation of the digestive organs, as in diarrhœa, dysentery, inflammation of the bowels and their covering, in advanced pregnancy, in affections of the kidneys and bladder, and wherever more drastic purgatives might unduly irritate; united with Epsom salts in doses of half or three quarters of a pound each, its action is prompt and certain in cattle, and for young calves it is the best of purgatives; it is a safe and easy purge for dogs and pigs.

Doses of the seeds to the dog or pig 6 or 8; of the oil for horses and cattle about a pint, for sheep and pigs 2 to 4 ounces, for dogs 1 to 2 ounces, and for cats 1 ounce.

OIL OF TURPENTINE.

The oil of turpentine, regarded by some as a sovereign remedy, is a most dangerous drug, and frequently aggravates the very disease it is intended to cure. The writer having seen several cases in which it produced death, he cannot recommend its internal employment to the uninitiated.

For external purposes it is often usefully applied with linseed oil, soap liniment, ammonia, &c.

TAR.

Tar is a stimulant, diuretic, diaphoretic, expectorant, and vermicide. It is not much employed internally. Externally it is, however, a capital agent in many complaints; for example, it is a good dressing for thrush and canker of the horse's foot, being used either alone or along with blue stone, sulphuric or nitric acids; mixed with fatty matters in equal parts, it is a splendid stopping for the foot, and keeps the hoof moist and soft, and also stimulates the horn. It is also of special service in foot-rot in sheep, and has the advantage of stimulating and deodorising noisome textures, and pre-

E

venting attacks of flies. It dries up grease in horses, and checks ringworm in calves. It is used for securing wounds, binding up broken horns, and making adhesive plasters.

Oil of tar is sometimes used instead of oil of turpentine. It is an admirable antiseptic, cures mange and scab, and destroys other parasites. It is often added to sheep dips, but has the disadvantage of discolouring the wool.

Pitch is also used in veterinary practice as a mild stimulant in such diseases of the horse's foot as thrush, canker, sandcrack, also in foot-rot in sheep, and for giving adhesiveness to plasters ; whilst its fumes are occasionally liberated by inserting a red hot poker into an iron pot containing the pitch, for the purpose of disinfecting surrounding premises.

TOBACCO.

The dried leaves of Virginian tobacco, *Nicotiana Tabacum.* Cultivated in America, B.P. Tobacco derives its name from the instrument used by the American aborigines for smoking the leaf, from the island of Tobago, or from the town of Tobasco in New Spain. It appears to have been cultivated from time immemorial by the natives of America, and is still largely grown about the great River Orinoco, in the United States, and in many parts of the world. It was unknown in the Old World, at all events in Europe, until after the discoveries of Columbus, and was first introduced into England by Sir Francis Drake in 1586.

Like alcohol, it has its admirers and inveterate opponents. A good deal can be said by both sides upon the question, but we have no desire to enter the controversy, further than stating its actual properties, actions, and uses. Of course, like every other article of consumption, it can be, and often is, abused by continued over-doses.

The *Nicotiana Tabacum*, which yields the Virginian and several important commercial tobaccos, is an herbaceous plant, with a branching fibrous root, a tall annual stem, funnel-shaped rose-coloured flowers, and large, moist, clammy, brown leaves mottled with yellow spots, covered with glandular hairs, and distinguished by a strong peculiar narcotic odour, and a nauseous, bitter, acrid taste. The leaves readily communicate their properties to hot water and alcohol.

The plant is cut down in August, and the leaves dried, twisted, and carefully packed with great compression in hogsheads. For many purposes the leaf is fermented, so as to destroy resinous and albuminous matters, which, when smoked, give rise to oils and unpleasant products. Sugar and liquorice are added to give mellowness and pliability. The several tobaccos of the shops owe their peculiarities chiefly to the manner in which they are prepared for sale; the manufactured Virginian, being the strongest, is generally prepared for medicinal purposes.

Snuff is prepared by cutting tobacco into small pieces, piling it into heaps, and pouring water over it to encourage fermentation. These heaps heat and evolve ammonia; the process continues during one to three months, according to the sort of snuff required, the fermented product is then ground and sifted.

Commercial tobacco contains about 12 per cent. of moisture, 20 to 25 of lignin, and nearly the same amount of inorganic matters, chiefly salts of potash and lime. The active principle is nicotine or nicotia, a colourless volatile, inflammable, oily alkaloid, with an acrid odour and taste. It is present in all parts of the plant, and constitutes from 6 to 7 per cent. of the dried leaf, and is soluble in water, alcohol, ether, and the fixed and volatile oils. It is an energetic poison, almost as potent as prussic acid.

Hertwig carefully investigated the action of tobacco on

the lower animals. He gave to horses ½ to 1 ounce of the
powered leaves, and found that the pulse was lowered three
to ten beats per minute, and became irregular and
intermittent, whilst a repetition of such doses increased the
evacuation of excreta and urine. Large doses injected into
the veins accelerated the pulse, increased the action of the
bowels and kidneys, and made the animal generally irritable
and restless. Two ounces of powdered tobacco, in 1½ lb.
of water, given in divided doses, but within two and a half
hours, to a healthy middle aged cow, heightened the skin
temperature, raised the pulse from 65 to 70, caused
quickened, but somewhat oppressed breathing, coldness
of the horns, ears, and extremities, dilatation of the pupil,
copious perspiration continuing all night. Next day the
animal remained dull; but by the third day she was all
right.

An ox consumed about 4 lbs. of tobacco leaves, and
speedily became very restless, ground his teeth, and groaned;
lay with outstretched limbs and distended stomach, passed
quantities of thin fœtid excreta, and died in eleven hours in
convulsions.

Its Medicinal Properties.—As a muscle relaxer, tobacco
acts very beneficially in colic, contraction of the neck of the
bladder, and occasionally in strangulated hernia. Impacted
colon and obstinate torpidity of the bowels, whether from
lead poisoning or other conditions, are usually relieved by
tobacco, given in the form of smoke clysters. One or two
drops of nicotine (equivalent to about 1 drachm of Virginian
tobacco) are given at intervals of two hours; it allays
the spasm of locked jaw in man, and a decoction applied
directly to the muscles has also afforded great relief. But,
as with other remedies in this disease, it is unfortunately
only temporarily removed. However, when administered,
it poisons intestinal worms; and diluted solutions, thrown

into the rectum as an injection, proves fatal to ascarides lodging there; but for any such purposes it must be used with extreme caution.

.. Externally, it is employed to kill the insect of mange in horses and dogs, and of scab in sheep; while it is fatal to lice, fleas, and ticks. Strong solutions liberally applied are liable to produce nausea, fainting, and even death; but 1 part to 30 of water can be used with perfect safety.

A useful dip for sheep, which is effectual in destroying ticks, warding off flies, and not injurious to the colour of the fleece, is made as follows :—1 lb. each of tobacco, sulphur, potashes, and soft soap, dissolved in 30 gallons of water. For such purposes the tobacco is previously boiled for ten or fifteen minutes in 2 quarts of water, and the decoction mixed with the other ingredients. These quantities are sufficient to dip thirty lambs, or a score of big sheep. For the destruction of scab, double the amount of tobacco may be cautiously employed.

Doses.—The large animals take from 1 to 2 drachms, sheep 10 to 20 grains, dogs 5 to 10 grains, dissolved in hot water.

As an antispasmodic laxative injection the smoke is preferable to the infusion, and is commonly given by filling a syringe with smoke drawn from an ordinary clay pipe. Three or four syringefuls are repeated at intervals of an hour as required.

For external applications, or for enema, the infusion is made by boiling or digesting 1 or 2 drachms of tobacco with a pint of hot water; stronger solutions require to be used with great caution, more especially if swallowed, injected into the rectum, or placed in contact with a broken and abraded skin surface. A single drop of nicotine destroys rabbits and small dogs in five minutes.

LIME.

When limestone, chalk, or marble, or any form of calcium carbonate, is mixed with coal, and thoroughly burned, its carbonic acid is driven off, and the metallic oxide or quicklime is left.

It occurs in greyish white irregular masses or lumps, has an alkaline-caustic taste, and a great affinity for water. It combines with 24 per cent. of water, giving off much heat, and forming the hydrate or slaked lime. The presence of sugar increases fully twelve times the solubility of lime in water.

Lime water is prepared by slaking a small quantity of freshly burned lime, agitating it briskly with a large quantity of water, allowing the undissolved matter to subside, and pouring off the clear solution. It is colourless, has an alkaline taste and reaction, and unites with oils to form soaps. As it readily absorbs carbonic acid, it should be kept in closely stoppered bottles.

Its Action and Uses.—Lime is irritant, corrosive, desiccant, and antacid. It is a natural constituent of the animal textures, in which it probably occurs mainly in combination with phosphoric and carbonic acids. Being present in most articles of diet, extra supplies are seldom required. When swallowed there is probably deposited on the gastric mucous membrane a film of carbonate, which is dissolved by hydrochloric acid, slowly absorbed as chloride, reconverted in the blood into carbonate, held in solution by the free carbonic acid, and ultimately excreted by the kidneys. Its effects are chiefly local.

Lime, especially when unslaked, and in contact with the mucous and abraded skin surfaces, attracts water, forms a coating of carbonate, and in large amounts irritates and corrodes.

Slaked lime and water are used as antacids in indigestion, hoven, and diarrhœa, especially among cattle. One-fourth or one-sixth of lime water given with their milk often prevents indigestion, flatulence, and diarrhœa amongst young calves, probably by counteracting undue acidity and the coagulation of the milk in large tough lumps. When acidity occurs with constipation, the bicarbonate of soda is substituted for the lime in the proportion of a drachm to the pint of milk; and even where there is no indigestion, lime water is often serviceable in ill-thriving calves and lambs. It is occasionally given as an antidote in poisoning by the mineral acids. By itself, and when turpentine is conjoined with it, it destroys bronchial filaria, often so troublesome in calves and lambs; and as an injection brings away the worms that inhabit the lower bowels. Mixed with oil or glycerin, it checks the discharge and abates the itching of eczema. But in such cases zinc preparations are as a rule more efficient. Scalds and burns are successfully treated with carron oil, which is simply lime water mixed with an equal quantity of linseed oil.

In powder and solution it is used for cleansing and deodorising foul stables, cow-sheds, and piggeries.

Doses.—Horses and cattle take 1 to 2 drachms of the quicklime, sheep 20 to 30 grains, dogs 5 to 10 grains. Of lime water, the larger animals require 4 to 5 ounces, and the smaller 2 drachms to 1 ounce, given with glycerine, oil, or, better still, milk. Two ounces each of lime water and gentian infusion, repeated twice or thrice daily, checks diarrhœa when it attacks feeble calves, and half the above quantity answers for sheep. For calves and dogs a useful antacid and stomachic is as follows:—2 ounces of slaked lime, 2 ounces of sugar rubbed down, then put into a bottle containing a pint of water, shaking the ingredients well together, and separating the clear solution, when the sediment has settled to the bottom. This is given diluted according to convenience.

CARBONATE OF LIME or CHALK.

This occurs in various forms of calcareous spar, limestone, marble, and chalk. Chalk is of especial medicinal importance, and abounds in the south of England. Prepared chalk is a dull, white, earthy, crystalline powder, tasteless, but adheres to the tongue, owing to its porosity and its affinity for water. It is a constituent of the bones of animals, of shells, and of coral. It is the common source of hardness in drinking waters, which when pure hold about 2 grains dissolved in the gallon. Sixteen grains are sometimes found in very hard waters, and when heated, the deposit corrodes the kettles and boilers.

Chalk is the cheapest and most convenient of antacids, and is much employed for all the domesticated animals in the treatment of indigestion, chronic diarrhœa and dysentery. It resembles lime, but is far less local in its effects. Its action extends throughout the whole length of the alimentary canal, neutralising acidity, absorbing irritants, and, during its absorption, leaving a film of lime salt which protects the intestinal surfaces. Thus excessive and faulty secretion is diminished. In a dry and finely divided state, it is used as a desiccant for external wounds and skin irritation, absorbing irritating discharges, and protecting from the air the host of ferments suspended in it.

Doses.—Horses take 1 to 2 ounces, cattle 2 to 4 ounces, sheep 2 to 4 drachms, pigs 1 to 2 drachms, dogs 8 to 12 grains, given in milk or gruel. When given often, care should be taken to keep the bowels open, in order to prevent it accumulating in the intestines.

MUSTARD.

The seeds of the *Sinapis nigra* and *Sinapis alba;* also the mixed seeds reduced to powder, B.P.

The mustard flour of the shops, according to information given to Sir Robert Christison by an English manufacturer, is made in the following manner :—" Two bushels of black and three of white seed yield, when ground, 145 lbs. of flour, which, to diminish the pungency and improve the colour, is mixed with 56 lbs. of wheat flour and 2 lbs. of turmeric, and the acrimony is restored without the pungency by the addition of 1 lb. of Chili pods and ½ lb. of ginger. Black seed alone, it is added, would be too pungent for use at table. Wild mustard seed is sometimes substituted for the black species, if the latter be scarce. Some manufacturers remove the fixed oil from both the white and black seed, by means of expression, before making them into mustard flour with the other ingredients, and the acrimony of the product is thus increased."

The Action and Uses of Mustard.—Unbruised mustard seeds have little effect when swallowed, probably because they are only partially and gradually digested. The flour, however, in large doses, is an irritant; in medicinal doses, it is a stomachic, carminative, and stimulant. A dessert spoonful dissolved in several ounces of water, and given to the dog or cat, causes vomiting. It is slightly laxative and diuretic, allied to horse radish and peppers, but is rarely given internally.

As an external irritant, it is in everyday use as a rubefacient, vesicant, or suppurant. A paste made with water and rubbed into the skin, within fifteen minutes causes redness, heat, and tenderness. Applied in larger quantity, or with smarter friction, the epidermis, after three or four hours, is separated from the true skin by effusion of serum, the vesicles run into globes of water, which subsequently burst and suppurate, the surrounding parts being swollen. The skin, however, generally heals in a week's time. Occasionally it may occur from repeated, prolonged, or injudicious use, as in irritable states of the skin there ensue active inflammation, sloughing, and destruction of the hair roots. Compared with cantharides, mustard is more prompt, but less permanent. It is used to control functional disturbance rather than to repair structural damage; it causes less exudation of serum, but more swelling of surrounding parts. Applied repeatedly, especially to the extremities of the horse, it is more liable to effect the skin deeply, and hence produce sloughing; and, unlike cantharides, it has no tendency to act upon the kidneys. It is almost as prompt, and is more manageable than boiling water. For horses it is less irritating and burning than oil of turpentine. It is not so severe or so apt to produce suppuration as euphorbium or croton oil. For cattle, mustard is an excellent blister, often acting promptly on their thick and insensible hides, when other agents have slight or tardy effects, and seldom causing injury or blemishing. For sheep and dogs it is ever useful, especially when applied, as it should always be, in moderate quantities, and for a short period.

For all the domestic animals suffering from colds, sore throats, bronchitis, pneumonia, and pleurisy, mustard in the early stages very seldom fails to lessen pain, and relieve the difficult breathing. It is most effectual when rubbed in over a considerable area, immediately external to the congested, painful, or inflamed parts, and in about twenty minutes washed off, and in an hour or two, if required, again reapplied. In acute indigestion, in colic, and typhoid fever, especially among horses, repeated dressings of mustard often afford relief. In phlebitis a smart blister reduces inflam-

mation, and hastens absorption of exudate. It is of service
in chronic rheumatism, particularly among cattle in the
second stages of inflammation of joints and tendons, in en-
largement of glands, and as a stimulant in chronic skin
diseases.

Blisters applied over the chest or belly, or below the
knees and hocks, especially when the limbs are cold, arouse
vitality and overcome congestion in the latter stages of
pneumonia and typhoid fever, in parturient apoplexy of
cattle, and in poisoning by narcotics. Conjoined with
stimulants, it is rubbed over the region of the heart to
counteract syncope. Applied over the kidneys, it promotes
diuresis. It is sometimes used to determine the secretion
of pus, for maintaining or increasing the effects of cantharides or mercury biniodide ointment. In horses, however,
considerable caution is necessary in applying one irritant
after another. Neither mustard nor any other blister should
be applied where the part is deeply inflamed, owing to the
danger of sloughing.

Doses.—As a mild stimulant, to the horse 4 to 6 drachms,
cattle ½ to 1 ounce, sheep and pigs 1 to 2 drachms, dogs 10
to 20 grains. The best mustard should always be selected,
and made into a paste with water, and well rubbed into the
parts affected, or over the region of the affected parts. If
great activity is required, it is better to mix with spirits or
vinegar. The volatile oil, prepared by distilling with water
the seeds of black mustard after the expression of the fixed
oil, is a prompt and powerful vesicant. Two drachms rubbed
into the skin of a dog caused immediate irritation, with the
speedy formation of large vesicles, surrounded by inflammatory swelling.

MYRRH.

Gum resinous exudation from the stem of *Balsamodendron myrrha*, collected in Arabia Felix and Abyssinia, B.P.

Myrrh is imported from the coasts of the Red Sea, chiefly by way of Bombay. It has been used from the earliest times in making incense, perfumes, holy oils, and unguents for embalming. It is at first of an oily consistency and a yellow white colour, but gradually becomes solid like gum, and of a brown-red hue. The best sorts, generally termed Turkey myrrh, are met with in irregular-shaped semitranslucent red-brown tears, which deepen in colour when breathed on. They are of variable size, brittle, and easy powdered; their fracture is irregular, shining, oily, and occasionally dotted with opaque white markings. Myrrh has a slightly bitter acrid taste, but an agreeable aromatic odour.

Administered internally, it is a bitter stomachic, feeble tonic and stimulant; externally it is a stimulant and astringent. It gently stimulates the digestive mucous membrane, and thereby excites secretion, bearing in this respect some resemblance to copaiva. It differs from the turpentines and balsams in possessing tonic properties; it is less stimulant and antispasmodic than the fœtid gum resins. It is sometimes given in indigestion, in chronic cough, and other mucous discharges; but its principal value in the Veterinary science is that it is a stimulant and antiseptic for wounds, applied in the form of tincture of myrrh.

Doses.—Horses and cattle take 2 drachms, sheep and pigs ½ to 1 drachm, dogs 10 to 20 grains, repeated several times during the day.

NITRIC ACID.

The strongest acid of commerce contains 85 per cent. of real nitric acid. The strongest acid of the Pharmacopœia contains 60 per cent. of anhydrous acid. On the large scale the commercial acid is prepared in iron retorts, from seven parts of sodium nitrate and four of sulphuric acid.

Nitric acid in tolerably concentrated solution is colourless; it emits pungent, corrosive suffocating fumes; possesses an intensely sour taste ; oxidises, corrodes, and dissolves many organic substances; dropped on the skin, it produces a yellow stain, deepened in colour by alkalies, and removed only by the wearing down of the part. It has great affinity for water. In imperfectly stoppered bottles it soon increases in quantity and diminishes in strength; diluted with water, it evolves great heat. Medicinally, it is a stimulant, and exerts antiseptic effects on relaxed and ulcerated conditions of the mouth and fauces. It diminishes the alkalinity of the blood, favours the flow of saliva and other alkaline secretions, and thus abates thirst and proves refrigerant. But large doses interfere with secretion of the gastric juice. It is sometimes prescribed in atonic dyspepsia, and where there is excessive fermentation, but in such cases it is not so suitable as hydrochloric acid. It is administered to cattle and sheep in chronic enlargements and fatty degeneration of the liver, in typhoid fever in horses, and alternated with arsenic in inveterate mange, eczema, and farcy. As a caustic, nitric acid is used for the removal of warts, fungous, and malignant growths, which cannot be removed with the knife, and for dissolving the hardened scurf which accumulates in neglected cases of scab and mange. It destroys the virus lodged in poisoned wounds, excites a healthy action, and removes the disagreeable odour from caries, foul and foot rot, and arrests spreading sloughing sores. When it is necessary to apply it to a sore, wrap a piece of tow round a splinter of wood, and well oil the parts surrounding the sore before you use the caustic. It is serviceable in abating the itching of nettlerash.

Excessive sweating in horses during exertion or sickness is often checked by sponging the skin with a dilute solution of the acid. Dissolved in 80 or 100 parts of water, it greatly relieves the tenderness and tension of piles in dogs.

Doses of the Diluted Medicinal Acid.—Horses or cattle take 1 to 2 drachms, sheep and pigs 10 to 20 drops, dogs 2 to 10 drops. It must be very largely diluted with water or other bland fluids, and is often combined with bitters.

For external application, a drachm of strong acid to the pint of water suffices for all purposes, escharotic excluded. A paste made with sulphur and lard is also in use for removing warts, destroying acari, and stimulating the skin in scab and mange.

WATER.

Two volumes of hydrogen and one of oxygen in the presence of a light, or an electric spark, unite with explosive force, yielding two volumes of gaseous water or steam. It exists in the solid, liquid, and gaseous forms. A cubic inch of water becomes a cubic foot of steam. When solid ice melts, heat is absorbed or becomes latent. When the liquid water boils or gives off gas still more heat is absorbed.

Organic matters when present, especially in river and marsh waters, cause them to spoil rapidly, and occasionally produce diarrhœa and dysentery. In suspension also occur such dangerous impurities as the germs of various catching

,diseases, and the ova of parasites. The solid constituents of
drinking waters vary greatly.

Glasgow has from Loch Katrine the purest water supply
of any large city in the world, containing only ¾ grain of
organic matter, and 1½ grain of inorganic matters to the
gallon.

The London water contains 3 grains of organic, and 16
grains of inorganic matters to the gallon.

Various plans are adopted for the purifying of water.
Subsidence and decantation get rid of grosser mechanical
particles. Filtration through sand, charcoal, or gravel
removes organic and organised impurities. Oxidation gradually destroys disagreeable or dangerous defilements, hence a
running stream contaminated even by sewage, a few hundred
yards lower down may again become clear and wholesome.
Boiling kills vegetable and animal matters, drives off
carbonic anhydride, and thus throws down calcium carbonate,
the cause of temporary hardness.

The Action and Uses of Water.—It is nutrient, diluent,
evacuant, and detergent. Hot water is an auxiliary emetic,
cathartic, and diaphoretic ; topically, it is emollient and
anodyne, and at still higher temperature is an active energetic irritant.

Cold water is refrigerant and tonic. It is applied to
wounds and burns in the form of the familiar water dressing,
and at low temperature it abstracts heat, and antagonises
local congestion and hæmorrhage.

Water is an unfailing constituent of all living tissues, and
is essential to the support of animal life. It constitutes a
large proportion of every kind of food, rendering it more
easily digested and assimulated. It supplies the loss of fluid
constantly taking place by the skin, lungs, and kidneys. It
promotes tissue change of form, and increases excretion.
Insufficient and excessive supplies of water are alike in-

jurious; but animals in health and in constant free access to water, rarely take more than nature demands. Excepting for a few hours previous to any great exertion, and when much over-heated and prostrated, it is unnecessary to restrict the water supplied to horses. In fact, in all well-constructed modern stables, water is always present at the horse's head, and when this plan is adopted he actually drinks less in the twenty-four hours than when he is allowed to slake his thirst three or four times a day. A moderate amount of water is essential for digestion; an excessive quantity injuriously dilutes the gastric juice. Horses when tired and hungry should have a few swallows of water, or, better still, a bucket of gruel, before feeding. A copious draught of water taken immediately after a rapidly eaten meal, hurries the imperfectly digested food too rapidly into the large intestines, and the result is too often colic, inflammation, &c. In febrile and inflammatory diseases water in moderation is a valuable medicine, and is perfectly safe, and a good deal more palatable and satisfying when given cold. Horses disposed to be greedy drinkers, and especially those that are damaged in the wind, should be supplied with small quantities and often, whilst further to relieve the thirst the food should be damped.

After a cathartic dose, and until the physic has ceased to operate, even moderate draughts of cold water in many cases cause griping. Calves and lambs feverish and purging, soon kill themselves if they have free access to water.

As a diluent, water mechanically relieves choking and coughing, dilutes corrosives and irritant poisons, assists the action of diaphoretics, diuretics, and purgatives; is mainly got rid of by the kidneys, lessening acridity of the urine, and augmenting its watery and solid parts. Tepid water is a convenient emetic for dogs and pigs; injected into the rectum, warm water allays irritability of the bowels and genital organs, and promotes the action of the bowels. Injections of cold water checks bleeding, produces general reaction, and occasionally expels worms. A good scrubbing with water and soap is the first course of treatment of mange or scab. It removes the scales, thereby allowing the dressing to go straight to its work.

Water is the important constituent of most emollients. Hot fomentations relieve tension, tenderness, and pain, and moisten, soften, and relax dry and irritable surfaces. Applied early, they control or prevent inflammatory congestion of bruises, strains, and severely contused wounds. Mainly by reflex action their application externally is simply invaluable, and soothes internal parts which have been irritated or inflamed. Thus fomentations allay the pain of colic, inflammation of the bowels, and all inflammatory affections situated within the chest. Steaming the head and throat in like manner, relieves strangles, colds, and sore throats. Such soothing vapours, medicated if need be by laudanum, belladonna, ether, vinegar, sulphurous acid, or alkaline hypochlorites, are readily evolved, from a well-made bran mash, placed in a roomy nose bag, or by holding the head over a bucketful of water, from which steam is driven off by plunging a hot iron into it at short intervals. Several folds of lint or tow saturated with hot water and covered with oiled silk, or gutta-percha cloth to retard evaporation, or a sheet of well-soaked spongia piline, are frequently substituted for fomentations and poultices, on account of lightness, cleanliness, and less tendency to sodden and injure adjacent parts.

Water nearly boiling is a prompt and powerful counter-irritant, and especially useful in cattle practice. It is laved over the affected part with a sponge or piece of flannel or soft rag. When applied to the chest or abdomen of cattle or horses, several folds of thick horse rugs or blankets are

F

often placed round the patient, and nearly boiling water poured amongst the folds of cloth from time to time. Thus the extensive counter-irritation, thus rapidly developed in careful hands, does not blemish, and frequently proves of great and permanent service in the first stages of pneumonia, pleurisy, colic, enteritis, peritonitis, and obstinate constipation both of horses and cattle. Cold water is a useful refrigerant. When the acute congestion, heat, and tenderness of bruises, strains, and other such injuries have been so far abated by hot fomentations, cold exerts wholesome, refrigerant, tonic, and constringing effects. Calico bandages, constantly kept wet, relieve chronic strains, jarfs, and wind galls in the legs of horses. The cold water treatment is also serviceable in broken knees, open joints, and circumscribed burns and scalds. Such wounds should not, however, be directly wetted, but kept scrupulously covered by folds of antiseptic wadding constantly wetted. Such continuous irrigation is readily affected through a small vulcanised india-rubber tube, brought from a supply tank on a higher level. Cold water similarly applied also keeps at low temperature the swabs around the coronets and feet of horses suffering from laminitis. Cold water dashed over the head and neck is a powerful stimulant, serviceable in megrims, sunstroke, phrenitis, convulsions, syncope, and the latter stages of puerperal apoplexy in cattle, as well as in poisoning with alcohol, chloroform, opium, and prussic acid, and for encouraging respiration in young animals that breathe tardily at birth. The shock is increased when very cold water is used, and when it falls on the patient from a height of everal feet. Ice in small fragments in a bag or bladder exerts more intense effects, and is valuable in inflamed and prolapsed uterus and rectum, and in those violent bleedings which occur at the time, or shortly after parturition. But care must be taken that vascular parts are not kept too long.

at such a reduced temperature as to interfere with their nutrition. Two parts of ice mixed with one of salt form a powerful freezing mixture of the temperature 4°, and are applied to prevent too sudden rise of temperature and gangrene in frost bite, to arrest circumscribed congestion and inflammation, to check bleeding, and to stop convulsions. Four or five minutes contact with the skin removes sensation, so that opening of abscesses, nerving, and such operations, can all be performed without pain; but for inducing local anæsthesia ether spray is preferable. Dr Chapman has taught that the ice bag applied along the back and the loins "not only exerts a sedative effect on the spinal cord, but also on those nerve centres which preside over the bloodvessels in all parts of the body it partially paralyses them." It appears to diminish muscular tension, sensibility, and secretion, and hence has been used in tetanus, chorea epilepsy, cramps, in neuralgic pain, and in inordinate discharges from the bowels and kidneys.

Baths, therefore, are important alike for the preservation of health and the cure of disease. Tepid baths cleanse the skin, promote perspiration, allay thirst, and are grateful to tired and heated horses. Hot baths stimulate the skin, incite perspiration, raise temperature, and, where long continued, quicken and enfeeble the pulse, retard oxidation, and impede electric currents through the nerves. They soothe animals subjected to severe muscular exertion, relieve colic and cramps, benefit chronic skin disorder, arrest colds and attacks of weed, promote the excretion of noxious matters, and thus prevent or alleviate rheumatism and various forms of blood poisoning.

Cold baths abstract heat or prevent its excessive formation, are tonic and stimulant; under proper control, they are useful in febrile cases, chorea, and convalescence from acute disease. As curative agents, they should rarely be continued for more than three or four minutes; whilst healthful reaction is encouraged by carefully drying, hand rubbing, clothing, and, if need be, stimulants. Vapour, Roman, or Turkish baths, when followed as they should be by cold effusion, combine most of the advantages of hot and cold baths. They are less depressing than the hot, and produce less nervous shock than the cold. They should not exceed the temperature of 120°. They promptly cleanse the skin, evoke perspiration, stimulate circulation, and increase both the destruction and construction of tissue. They are specially useful in chronic cough, dyspepsia, want of appetite, rheumatism, laminitis, in the shivering cold stage of fever, and in disorders depending upon blood contamination.

Where proper baths cannot be obtained many of their curative advantages are secured by rapidly sponging the animal with tepid, hot, or cold water. Noxious and irritable matters are removed from the skin, circulation is equalised, excessive heat reduced, and spasm counteracted. In febrile cases, either in horses or cattle, the temperature of the water at first should not be more than 80° or 85°, the sponging should not occupy more than three minutes; the animal should be dried and immediately clothed. Within two or three hours the process may be repeated, providing the temperature rises again.

From the foregoing remarks it will, therefore, be observed that water is a most invaluable curative and medicinal agent, alike admissible to all patients, from the highest scale down to the lowest of our domestic animals.

GENTIAN ROOT.

The dried root of *Gentiana lutea*, collected in the mountainous districts of Central and Southern Europe, B.P.

Gentian is the type of a pure and simple bitter, and is prescribed as a tonic and stomachic, promoting salivary and gastric secretions; and as a tonic it has been considered inferior to cinchona, but it is devoid of astringency. It, however, improves the appetite and general tone, and imparts a healthier condition to the stomach.

Among horses suffering from febrile colds, no combination is more useful than an ounce of gentian (powdered), 2 drachms of nitre, with 2 ounces of Epsom salts, dissolved in a pint of water, linseed tea, or ale, and repeated night and morning, in most inflammatory complaints. Such a remedy also proves serviceable after the first acute stage is passed. When the bowels are constipated and irregular, or febrile symptoms are insufficiently subdued, 2 drachms of aloes may be given with advantage in the mixture ; and when a more decided tonic effect is desired, iron sulphate is alternated with the gentian and salts. One ounce of gentian with 1 ounce of ether, or sweet spirits of nitre, given three or four times a day in a bottle of good ale, proves an excellent tonic, stomachic, and stimulant in influenza, and many other debilitating diseases; and not only so, but it hastens the recovery from exhausting complaints, and possesses an almost magical effect in restoring jaded horses, or animals that are over-worked or suffering from loss of appetite or colds. In simple indigestion in young animals it is often combined with aromatics and antacids, such as half an ounce of gentian, ginger, and carbonate of soda, constitutes a very useful carminative and stomachic for horses or cattle, and may be made either into a ball with treacle, or into a drink with gruel and ale, and thus administered.

In relaxed and irritable bowels, especially in young animals, it is advantageously given with opium. By promoting a healthier state of the digestive organs, it prevents the development of worms, while its bitterness and slight laxative tendency often cause their expulsion. Its supposed utility in jaundice is principally due to the laxatives with which it is combined. For sheep, gentian is a most useful stomachic and bitter tonic, and when prescribed with salt arrests for a time the process of liver rot. Next after quinine, it is the best vegetable tonic for dogs prostrated by reducing disorders, such as distemper, &c. Like other tonics, it is not admissible in irritation of the intestines and in the early stages of acute inflammatory diseases. As an infusion, it is sometimes applied externally as a mild stimulant and antiseptic.

GINGER.

Zingiber, the scraped and dried rhizome, or underground stem, of *Zingiber officinale*, from plants cultivated in India, in the West Indies, and other countries, B.P.

Ginger is slightly irritant, aromatic, and stomachic. It stimulates the various mucous membranes with which it comes in contact. Blown into the nostrils, it promotes nasal discharge; chewed, it increases the flow of saliva; administered internally in repeated doses, it increases the gastric secretions, facilitates digestion, and checks the formation of flatus. Owing to these stomachic and carminative properties, as well as from its mild tonic effect, it proves serviceable during convalescence from debilitating diseases, especially when accompanied by atony of the digestive organs. It is, besides, a useful adjunct to many medicines, and is prescribed with tonics and stimulants; whilst, combined with purgatives, it diminishes their liability to nauseate and gripe, and also hastens their effect, therefore it is used for all domesticated animals to fulfil those purposes, and is especially adapted for cattle and sheep.

Doses.—Horses 4 drachms to 1 ounce, for cattle 1 to 3 ounces, sheep 1 to 2 drachms, pigs ½ to 1 drachm, and for dogs 10 to 30 grains. It is given in balls made up with treacle, &c., or in a draught, with ale, or hot water sweetened with sugar or treacle.

GLYCERIN.

A sweet principle obtained from fats and oils. It was first discovered by Scheele, in 1879, as a product in the manufacture of lead plaster. It occurs in small amount during the fermentation of sugar, and is obtained as a by-product from soap and stearine candle-making. It is viscid, colourless, and odourless ; has a sweet taste. Strongly heated in a capsule, it should, if pure, leave no residue. It burns with a luminous flame, evolving irritating vapours ; is freely soluble in water and alcohol ; is in itself an excellent solvent for vegetable acids and alkaloids ; takes up one-third of its weight of quinine sulphate, one-sixth of morphine muriate, and one-fifth of arsenic.

Its Medicinal Properties and Uses.—Glycerin is nutrient, demulcent, emollient, feebly antiseptic, and a convenient solvent for alkaloids, tannic and gallic acids.

It possesses some of the nutrient properties of cod-liver oil and other fats, and 2 or 3 drachms repeated twice or thrice daily have been administered to delicate dogs. It has been stated that 150 grains act like alcohol, and kill small dogs. Small doses are thrown off by the kidneys, large doses by the bowels, producing a laxative effect. It is a splendid application for cracked heels, mud fever, blistered or burned surfaces; in short, wherever the skin is irritable, dry, rough, or scurfy, its use is attended with good results. For tender or abraded surfaces, when undiluted, it is too heating and irritant, and is best used with oil, or water, or any bland fluid. It is a cleanly and useful application for many purposes, notably for sore mouths amongst lambs and calves. Greater antiseptic and astringent effects are secured, as is desirable, in bad cases of cracked heels, in harness galls, or indolent wounds, by the application of glycerin and carbolic acid; this is easily made by rubbing together in a mortar four parts of glycerin to one of carbolic acid. Another soothing and astringent dressing is made with equal parts of glycerin and Goulard's extract, diluting it as required with water. It is especially useful for preserving balls and masses in a soft sound state, and is convenient and palatable for giving disagreeable medicines to dogs, &c.

SODA AND ITS MEDICINAL COMPOUNDS.

The carbonate and bicarbonate of soda are antacids and alteratives. They are constituents of the blood and bile and serous fluids. Used as a medicine, they restore any deficiency of the soda salts. During their absorption, their continuance in the body, and their expulsion therefrom, they exert antacid properties; or, in other words, combine with lactic, uric, and other organic acids. They are also antidotes for poisoning by mineral and other acids. Beneficial effects result from their use in moderation in febrile attacks of rheumatism and irritation of the kidneys; but injurious and wasteful effects ensue from their continued abuse. Small doses increase the secretion of the gastric juice, and assist the emulsion and digestion of fats. Young calves, too exclusively fed on skim milk, and suffering from indigestion with constipation, are often relieved by dissolving 3 or 4 drachms of bicarbonate of soda in each meal of milk.

The carbonate and bicarbonate differ only in their degrees of action. They closely resemble the corresponding salts of potash, are, however, less penetrating and irritant, but contain a greater percentage of alkali, and possess a stronger neutralising power.

Doses.—Horses and cattle take 2 to 6 drachms, sheep and pigs 20 to 40 grains, dogs 10 to 20 grains. The bicarbonate is administered in exactly double these doses. It is best given in solution.

SOAPS.

Soaps are laxative, emetic, antacid, and diuretic. They are used externally as stimulants, lubricants, and detergents, and in pharmacy as excipients. They form convenient adjuncts to more active laxatives or diuretics; and some practitioners recommend their adoption in injections. But to this I most emphatically take objection, and for the following reasons:—In the first place, there are very few individuals who have not suffered painful irritation by allowing a small quantity of soap to enter the eyes while washing the face. That being so, it soon manifests its irritant effects upon the sensitive membrane of the eye. What must it be then when a thick soapy injection is ushered into contact with the mucous membranes of the bowels? I leave you to reflect upon the torture inflicted upon the helpless animal. But as a stimulant for bruises and strains, and to produce counter-irritation in cold and sore throats, 6 ounces of hard soap, cut small, and macerated with 6 ounces of dilute liquor ammoniæ, and 1 pint of proof spirit, and 1 pint of linseed oil, with 3 ounces of camphor added, is indeed valuable for the above complaints, well rubbed in. Soap also prevents access of air to burns or scalds, and by doing so greatly relieves irritation. It is also greatly employed for making up balls, liniments, and ointments.

SULPHATE OF ZINC

Is an irritant, sedative, emetic, antiseptic, astringent, and nerve tonic. It is used externally as an astringent, stimulant, antiseptic, and desiccant. As is often the case with other metallic poisons, several ounces are given to horses and cattle with impunity. Orfila discovered that 7½ drachms were vomited by dogs in a few seconds, but produced no lasting effects. Christison says "that 30 grains in solution, injected into the veins, depressed the action of the heart, and destroyed life in a few seconds." As a tonic, it resembles but is inferior to iron and the sulphate of copper; as an astringent, it is given along with opium in dysentery and diarrhœa, but is not so efficacious as sulphate of copper, &c.

Externally, it is greatly employed as a stimulant and astringent in oversecreting wounds, in foul ulcers, in inflammatory affections of the eyes, in chronic skin diseases, and in inflammation between the digits of the sheep's foot.

As an emetic for dogs and pigs, 8 to 15 grains (according to size) in 2 or 3 ounces of water. As an astringent, for horses or cattle 1 to 3 drachms, for sheep 10 to 20 grains, for dogs 2 to 5 grains are given, either in the solid or fluid state.

Externally, it is used in powder or solution. An ounce each of the sulphate of zinc and acetate of lead, dissolved in a quart of water, forms the well-known white lotion used in veterinary practice.

OXIDE OF ZINC.

It is seldom that this agent is prescribed internally. Externally, the oxide in powder is dusted as a desiccant, stimulant, and astringent, over chafed and irritated skin surfaces (for example, mud fever), and is also employed in solution or ointment. The B.P. process is made by mixing with gentle heat 80 grains of oxide of zinc and 1 ounce of lard.

ARECA NUT or BETEL NUT.

This is a most effectual vermifuge, particularly in dogs; and is alike fatal to both tape and round worms. Before its administration the bowels should be thoroughly cleared out by a purgative, and the areca nut given fasting. The worms thus starved are eager for anything, and thus they greedily swallow the poison prepared for them.

Half a nut, or about 60 grains of the powder, is sufficient for a pointer. The nut should always be given powdered either in milk or soup, and in a few hours the worms will be discharged. Sometimes it is advisable to follow up in a day or two a mild dose of castor oil with a little turpentine, so that worms that previously appeared immovable will very often be entirely removed.

Doses.—The dogs will take from 15 grains to 2 drachms, horses 4 to 6 drachms, given in milk, as worms are partial to it.

CATECHU

Comes from Bengal and Burmah in great masses, which weigh about 1 cwt., and made up of layers of small pieces of 2 to 4 ounces, enveloped in husks of rice. It is bitter and astringent, and dissolves in the mouth slowly. It owes its value for tanning and its medicinal properties to the presence of tannic acid, &c.

It is administered when the mucous discharges are excessive, and is conjoined with aromatics to remove flatulence, with opium to remove irritability, and with chalk, magnesia, or any alkali to counteract acidity. For diarrhœa the following is convenient:—Three ounces each of catechu, prepared chalk, and ginger, with 6 drachms of opium, made into a mass with treacle and linseed flour. This makes 6 doses for a horse, 4 for a cow, and 8 or 9 for a calf or sheep.

ENUMERATION OF THE DISEASES APPERTAINING TO THE LOWER ANIMALS.

LAMENESS.

Navicular Disease or Grogginess.—Inflammation of the navicular bone of the foot. Treatment :—Remove the shoes, place the patient in a loose box laid with clay; apply cold water continuously; failing that, blister smartly round the coronets, and rest. Should this course prove ineffectual, resort to neurotomy or nerving.

Laminitis or Founder.—If due to irritation and inflammation of the mucous membrane, the giving of purgatives are dangerous. But if brought on by indigestion, then give a dose of physic. If fever is great, combat it with aconite ; if pain is excessive, give opium; follow up with hot bran poultices when the shoes are removed. If the animal will not lie, throw him down; he will award you his best thanks by a sigh of relief. If there is acidity of the stomach, he will lick the wall. Give carbonate of soda, when acute symptoms subside; put on bar shoes, and apply cold water constantly. Diet to consist of bran mashes, with very little fodder. If after three weeks he still goes lame, blister smartly round the coronet, and rest.

Seedy Toe.—If due to the clip of the shoe, remove the cause, separate the healthy from the unhealthy; failing that, strip the part entirely, blister round the coronet, and rest.

Thrush.—Give a dose of physic, poultice for a day or so, keep clean, and dress with sulphate of copper or calomel.

Canker.—Strip the sole and other portions of wall where the cancer is situated, apply tincture of iron on tow, and bandage up the foot. If there is fungus, it must be removed by caustic; but be cautious not to injure healthy parts. When the parts become covered with horn, apply gentle pressure.

Carbuncle.—Inflammatory process affecting coronary substance. The point of inflammation must be severed by a crucial incision, then dress with astringents, inject strong carbolic acid to alter the state of inflammation. Afterwards poultice; keep the bowels open; if fevered, give aconite. Dress with sulphate of copper, and blister should inflammation involve whole substance.

Villitis.—Inflammation of the coronary band. Causes: —Excessive work, bad shoeing, high caulking. Cold water applications, thin heeled bar shoe, blister, rest.

Quitter.—Causes:—Treads, grease, pricks, festering corns.

Canker or Gathering Nails.—Remove the shoe, search for the cause, and remove it. Thin the horn with rasp and knife, poultice thoroughly, and inject solution of corrosive sublimate, setons, &c.

Ring Bone is of two kinds, true and false. Apply soothing remedies, hot poultices; dose of physic. Secondary treatment:—Put on bar shoe and should the disease continue, fire and blister, and rest three months.

Side Bone.—Analogous to ring bone, excepting seat. Treatment the same as for ring bone.

Corns, &c.—They are produced by careless shoeing or tight shoes. Remove the pressure, not by mutilating the foot, but by springing the shoe. Pricks and other injuries must be treated as follows:—Removal of the cause, and the effect will cease; second, soothing remedies; and lastly, if obstinate, counter-irritation and rest.

Rachitis.—Due to deficiency of earthy matter in bones. Give an excess of lime water, put splints on the limbs, and bandage not too tight.

Splints.—Reduce inflammation by fomentations, give a purgative, and blister; failing this, separate the exostosis from the bone with the knife. If lameness continues, counter-irritation and rest.

Spavin.—Bog and bone. Subdue inflammation; if he stands on toe, apply high-heeled shoe. Fire, blister, and rest.

Sprains are violence inflicted upon soft structures. If of the back, dose of physic; if restless and fighting and feverish, opium, aconite, soothing applications, liberal nourishment, and rest.

Shoulder Slip.—Hot fomentations for three or four days and nights. Afterwards blister, and allow complete repose.

Capped Elbow.—Cause:—Lying down on shoe. Prevention:—Put on pads, dissect out tumour, and apply carbolic oil.

Inflammation of Knee-Joint or Carpitis.—Treatment unsatisfactory; however, reduce the inflammatory process, counter-irritation, and rest.

Broken Knees.—Cleanse thoroughly from all dirt, and apply carbolic oil.

Rupture or Sprain of the Suspensory Ligament, or (in racing parlance) Break-down.—Fill up the hollow of the pastern with pads tightly wedged in. Bandage, and treat with cold water.

Shivering String Halt.—Nervous disease. Very little control over it; in fact, incurable.

Hip-Joint Lameness.—One leg appears longer than the other; whole limb stiff and straight. Reduce inflammation, afterwards fire and blister.

Disease of the Stifle Joint.—Absolute repose; insert a seton on each side, and allow them to remain in for three weeks.

Capped Hock.—Difficult to get rid of; soothing remedies are the best. If the swelling is hard and hangs over, a seton run through may have good results.

Thorough Pin.—Railway horses used for shunting are very liable to it. Treatment very unsatisfactory.

Curb.—Treatment:—Counter-irritation and rest.

Chronic Grease.—Confined to hind extremities as a rule. Causes:—Want of exercise, bad state of the body; constitutional predisposing, intrinsic and extrinsic. Treatment:— Remove the grapes with hot fire-shovel (such as is used in the forge), afterwards poultice, and dress with astringents, or pure carbolic acid; dose of physic, &c.

DISEASES OF THE ORGANS OF RESPIRATION.

Common Cold or Catarrh.—Treatment:—Arrest it at once; delays are ever dangerous. The most effectual means are the soothing ones, hot cloths, steaming the head, and giving a little nitre in the water to drink.

Laryngitis.—Treatment the same as common cold.

Roaring—It is a vain imposture to offer specifics for the cure of this disease. All we can do is to avoid breeding from animals predisposed to it.

Bronchitis.—Inflammation of the bronchial tubes. Treatment :—Place animal in a cool atmosphere, clothe the body and keep legs warm. Do not give aloes on any account; give nitre half an ounce, and tartar emetic 1 drachm in water. If bowels are constipated, unload with injections, apply plenty of hot water to chest and sides, and support the patient's strength. When the first stage is passed, and weakness ensues, give stimulants, and continue hot cloths. When pulse runs up, give aconite. When these medicines fail, give powdered camphor and extract of belladonna, of each 1 drachm, every three hours. The disease may terminate in several ways, such as recovery, or congestion, or pneumonia, or in thick or broken wind. The latter every one knows by its deep hollow cough.

Congestion.—Brought on by exposure, impure air, rapid and over-work. Place in a comfortable box, give a stimulant as early as possible—whisky, if it is handy, with aconite, and the probability is he will be all right again in a little while. To be brief, the treatment is simply stimulants, sedatives, comfort, and pure air.

Pneumonia or Inflammation of the Lungs.—In health the lungs float in water; in disease things are the reverse. This disease is ushered in by a shivering fit, cold skin, with harsh staring coat. If patient gets over nine days, you may expect recovery. Treatment :—Clothe the body, and keep comfortable. Hot fomentations, aconite, nitre, sweet spirits of nitre, and careful nursing, stimulants, &c.

Pleurisy is a very painful complaint. We have much fever and colicky pains, breathing is quick and laboured. Treatment in every way similar to pneumonia.

INFLUENZA.

The term signifies *influence*, from the superstition that it was influenced by the stars. It is a very vague name, as it does not imply any particular pathological disease in the lower animals ; but in man the term is restricted to certain symptoms.

Symptoms.—Well-marked dulness ; no shivering fit ; the patient stands hanging his head in a corner. A cough is present, which is sore and painful, and at times almost choking. The breathing is accelerated, and in a day or two we have an early discharge of mucous from the nostrils, which is very distinctive of this disease. The mouth is hot and dry, the surface of the body is irregular in temperature, some parts being hot, others cold. We have a small, quick, weak pulse, with general weakness and sinking. Such are the symptoms when the surgeon is generally called in.

The next stage is, pulse running up to 80°, and small. Cough increased, lifting of the flanks, and a ridge running along the sides to the flank, the same as in pleurisy; extreme weakness. If you turn the patient round he staggers ; the mouth hot and dry. With these symptoms, we must come to the conclusion that there is typhoid pleurisy and typhoid bronchitis.

Treatment.—First a cool pleasant box, made comfortable; plenty of sweet straw. Clothe the patient, envelop him in hot blankets wrung out of boiling water, and give aconite and sweet spirits of nitre, with chlorate of potash in a pint of warm ale every six hours. But half of the cure lies in careful nursing and attention.

INFLUENZA OR DISTEMPER.

Symptoms.—When the animal is first seized, he is very stiff, the belly is tucked up, skin sticking to the ribs, or hidebound, and he rarely moves. Next day legs are swollen, also swelling under the abdomen, pulse quick and unnaturally weak ; legs, ears, and surface of the body cold ; and all the openings of the body are swollen.

G

Treatment.—Clothe the body, and make· the anim comfortable. Allow him to eat any sort of good nourishi food ; give mild diuretics followed up with mineral toni and, as a rule, recovery is the result.

THICK WIND

Often follows bronchitis, influenza, &c.

Treatment.—Sedatives may be given, but the best thi that can be done is to endeavour to palliate the disea: which is done chiefly by the liberal use of the best food small quantities, water given often, and little at a time.

BROKEN WIND

Resembles asthma in the human being. The symptoi are well-defined, and therefore the disease is quite famili to all those acquainted with horses.

Treatment.—Some assert it is curable. The writ declares it is not. You can arrest the appearance of blo ing, and mitigate the aggravated symptoms, but to cure a remove the cough you never can.

DISEASES OF THE HEART.

Common in the human being, but rare in the domest animals. However, functional disease of the heart is ofte witnessed where we have great debility, also in the disea called stomach staggers, in which both brain and hea frequently sympathise with the stomach.

DILATATION OF THE HEART.

This also seldom occurs.

TUBERCULOSIS AND CONSUMPTION.

It is estimated that one-sixth of the population of tl country dies from this disease alone.

Treatment.—With regard to this, we cannot boast much, although every paper is crowded with specifics. If in the early stage we can lay on a little fat with good oil cake, &c., by all means do so. Still the great trouble and expense is not adequately rewarded.

HICCOUGH, HICCUP, OR SPASM OF THE DIAPHRAGM,

Often caused by overloaded stomach.

Treatment.—A drink of cold water will often cure it, or 2 ounces of sweet spirits of nitre and 2 ounces of opium tincture.

PALPITATION OF THE HEART.

The sound is often heard at some distance from the patient. Treat with tincture of opium.

DISEASES OF THE STOMACH, &c.

Indigestion.—The term is often vaguely employed. It is applied to an imperfect action in the stomach and intestines, but, correctly, should be limited, as in man, to affections appertaining to the stomach alone. Treatment:—Change diet totally, give a dose of physic, and carbonate of soda in the drinking water.

Vomiting—A horse can only vomit through his nose. It is produced by irritation, &c.; give tonics and sedatives.

Stomach or Sleepy Staggers.—It is called stomach staggers, because it depends upon a filling of the stomach tight, and sleepy staggers because the brain sympathises with it. Treatment :—A good dose of physic, to be followed up with stimulants if pulse is low. Give carbonate of soda to neutralise the gas, hand wisp belly, &c.

Grass Staggers.—Symptoms similar to what are seen in stomach staggers, only modified. Treatment:—Get rid of the impacted food by a dose of physic at once; and, as a rule, there is recovery.

Gastritis, or Inflammation of the Stomach.—There can not be the least doubt but that this organ is liable to inflan mation, although in the horse it is generally accompanie by inflammation of the bowels. In man, dog, and pig gastritis occurs as a distinct disease.

INFLAMMATION OF THE BOWELS OR ENTERITIS,

Destroys life in a few hours. Causes are various; th same that produced stomach staggers, &c., will produc enteritis.

Treatment.—Bleeding hastens death. Apply hot blanket give opium, aconite, nitre, powdered aloes, belladonna, &c.

After Consequences.—Death is the most common issu or it may terminate in acute founder. If so, keep feet coo attend to patient's comfort, and disturb him as little possible.

CONSTIPATION.

Millers' and bakers' horses are the most subject to du balls, or animals fed on dry food.

Treatment.—You must employ all the remedies that ca be resorted to; but eventually the ball causes death.

COLIC

Is of two kinds—one flatulent, the other spasmodic. Th latter is the most common, and is easily relieved. Its attac is sudden and without fever. In the first form the abdome becomes distended with gas; there is no cessation of th symptoms.

Treatment.—Give a dose of physic; this is the best saf guard, and one that should never be neglected; follow u with opium and sweet spirits of nitre; and, as a rule, yc have a favourable termination.

PERITONITIS

Is inflammation of the peritoneal membrane which lines the walls of the abdominal cavity, while it frequently extends to that which forms the external coat of the intestines. It occurs frequently in the cow.

Symptoms.—Fever and restlessness, animal strikes his abdomen with his feet, seldom lies down, being afraid; quick strong pulse, breathing quick and short, straining and stretching, and intense pain.

Treatment.—Hot fomentations applied without delay, and continued without stoppage. Give a good dose of physic; it relieves the loaded intestines. Follow up with opium, aconite, and nitre. Injections should be given, and persist with the hot water.

LIVER DISEASE.

Common in sheep; give calomel 5 grains, powdered opium 4 grains, in gruel once a day.

STRANGLES.

Symptoms.—Ushered in by a catarrhal affection, sore throat, cough, and a discharge from the nose. In three or four days great difficulty in swallowing. Swelling is seen underneath the jaw. In true strangles the inflammation affects the whole of the glands.

Treatment.—If season will allow, give green food; liberal diet, which must be good; regulate the bowels by the food; apply hot fomentations; give nitre in the water. After the abscess has burst, should a cough remain, apply a smart blister.

GLANDERS.

A dangerous and fatal disease, alike to horse and man.

Treatment.—There is none.

PURPURA HÆMORRHAGICA.

This name literally means purple blood. It follow diseases such as strangles, bronchitis, diabetis, excessiv purgation, &c.; or, in fact, any disease that tends to debili tate the system.

Symptoms.—Swelling of some part of the body, legs first as a rule, beginning below, and rapidly involving whol limb. The head is also swollen, often blood trickling fron the nose and eyes; blotches are seen inside the nostrils When the disease advances the legs crack, and blood ooze through—not true blood, however.

Treatment.—The great object is to keep patient comfort able and supplied with cool air, while as much nourishmen must be got into him as you possibly can get him to swallow Give tonics, whether he is inclined to take them or not especially iron. If the debility is great, give stimu lants. Turpentine is recommended in this disease, but eschew it.

ERYSIPELAS

Is a disease of frequent occurrence in man; and a curiou one it is. It has several names, such as the "Rose" anc "St Anthony's Fire." It is more frequent in the cow thai horse.

Symptoms.—Begins below and extends upwards, usuall the legs or head. The skin is thickened and raised up, anc feels brawny, like the skin of swine; that is, hard, rough and rigid.

Treatment.—Open the bowels, which must be done b diet. Give diuretics and vegetable tonics; iron does no seem so needful in this case. If the season permit, giv green food. Local treatment: hot applications, and th liberal use of oil or lard to the swollen parts.

SCARLATINA.

Many people are greatly alarmed at the existence of scarlet fever, and watch its inroads with horror, while the scarlatina creates no fear or anxiety. It may be a happy delusion, but it would be interesting to many to know that the distinction is in name only—the different terms signify one and the same disease.

Treatment.—Get the bowels opened; and then with careful nursing, and a little nitre in the drinking water, they soon recover.

DISEASES OF THE NERVOUS SYSTEM.

The diseases affecting this system, in number if not in nature, are much fewer than what is observed in man, especially in the horse.

TETANUS OR LOCKED JAW.

So called because it locks or fastens together the jaws. It is a disease of the nervous system, inducing a continual spasmodic contraction of all the voluntary and most of the involuntary muscles of the body. Sometimes the head and neck is only affected; when so it is called trismus.

It occurs under two distinct forms—traumatic and idiopathic.

Causes.—Wounds or irritation of the spinal cord.

Treatment.—It is treated in all sorts of ways, but the best is to support the system, keep the animal perfectly quiet, and get the bowels open.

MEGRIMS

Consist in temporary loss of voluntary power and motion.

Symptoms.—Animal shakes its head, then looks at its flanks, and falls down.

Treatment.—Allow a roomy collar and food that will no constipate the bowels; give an occasional dose of medicine.

CHOREA OR ST VITUS'S DANCE.

More frequent in the dog than any other domestic animal.

Men stagger although they are not drunk, and look idiotic though they are not.

Treatment.—There is none.

STRINGHALT.

Involuntary twitching of one or both hind legs.
Treatment.—None.

PHRENITIS OR MAD STAGGERS.

This is inflammation of the brain and its coverings.

Treatment.—Give a large dose of purgative medicine; shut the animal up in a dark place, apply cold water cloths to head, and give injections to hasten the action of medicine. Get all out you can in the shape of food, and allow nothing to be eaten until the symptoms have passed away. Never apply blisters; they excite, and do not allay irritation.

TUMOURS

Are of two classes—non-malignant and malignant. The first are composed of materials already existing in the body. They do not destroy, poison, or spread to surrounding tissues, having no tendency to reproduction; and when thoroughly removed, the animal recovers. They may destroy life, but this is not due to poisoning of the system, but from mechanical inconvenience.

Malignant tumours, on the other hand, such as cancer, do destroy, change, and alter the textures surrounding them into their own textures, while they poison the whole system

They cause great pain and irritation, and when removed there is no guarantee that they will not be reproduced in same or another site.

Treatment.—For all tumours that can be removed with the knife, do so, and afterwards apply carbolic dressings.

RINGWORM.

Treat with acetic acid or iodide 1 drachm, lard 1 ounce, mix and apply; and in bad cases, the iodide of silver.

WOUNDS.

Treatment.—Wash well, and remove all irritants; arrest the bleeding if any, and apply carbolic dressing.

BURNS AND SCALDS.

Treatment.—Subdue the fever by giving a laxative and aconite; apply lime water and linseed oil equal parts, and protect from the atmosphere.

TYMPANITIS, HOVEN, OR DISTENTION OF THE FIRST STOMACH.

It is produced in cattle by the fermentation of food in this organ.

Symptoms.—Great distention of the stomach on the left side. If struck, it produces a hollow drum-like sound; nose poked out, breathing hurried as swelling increases, &c.

Treatment.—When the distention is great, insert the trocar at once, and liberate the gases; then gag the mouth. Give a dose of physic first, and change diet.

IMPACTION OF THE FIRST STOMACH.

Give an energetic purge, and repeat in twelve hours if necessary. Failing that, the side must be opened, and the impacted food removed with the hand; afterwards stitch up stomach and side. Be cautious with the diet, attend to comfort, and keep the patient as free from movement possible.

WHITE SCOUR IN CALVES—DIARRHŒA.

The cause is generally traceable to the food, which course must be changed. Then give castor oil or Epso salts and treacle, with stimulants. When the bowels a cleaned out, give a switched egg and a little port wine a1 ginger mixed, and attend to the general comfort.

INFLAMMATION OF THE UTERUS OR WOMB.

Mostly seen after calving.

Symptoms.—Pulse quick and wiry, breathing hurrie animal staggers, a blackened fluid is discharged, while tl patient moans repeatedly. It is very fatal in cows.

Treatment.—Inject tepid water into the womb at interval apply hot cloths wrung out of boiling water. Give digital: opium, nitre, and as recovery advances, good food a1 tonics.

PARTURIENT FEVER.

Generally ushered in two or three days after calving. Co gets restless, no inclination to eat, rumination suspende milk scarce, looks round to her flank, lies down, moar stretches out her head, attempts to rise, kicks at her bell great fever; ears, horns, and leg cold; pulse gets weak(the breathing quicker. Finally, she lies with head ere(retains senses to the last. Cows may labour for two three days, but the duration is generally about forty-eig hours, when death ends the scene.

Treatment.—An active purge at first, one that will ; straight to its work. Give the croton beans, the Epsom sal{ and the linseed oil, with 3 or 4 ounces of tincture of opiun give also injections, apply hot fomentations to the spine, mi often, and turn her often ; give powerful stimulants and linseed tea, and during convalescence attend to her comfort.

MAMMITIS OR INFLAMMATION OF THE UDDER.

Either produced by cold, wet, blows or kicks, or by carrying an overloaded udder.

Treatment.—Hot fomentations ; apply a strong bandage, make holes in it to fit the teats, then fasten the bandage over the loins. This gives immense relief. Give a dose of physic, and rub the udder with plenty of lard. If teats are blocked, insert a siphon ; should abscesses exist, open them, and dress with carbolic oil.

BLACK QUARTER, QUARTER ILL, BLACK LEG.

Cattle may appear all well at night, and in the morning you may find one dead. It is a very fatal and rapid disease.

Treatment.—This is a matter of considerable difficulty, owing to the rapidity of the disease. Purgatives at first, and keep the patient walking about. Insert setons into the dew-laps of all the stock, and allow these setons to remain for three weeks, moving them daily. This with change of food is the best prevention.

APHTHA OR THRUSH.

This consists of roundish, pearl-coloured vesicles confined to the lips, mouth, and intestinal canal, and generally terminates in curd-like sloughs.

Symptoms.—Swelling of the membrane that lines the mouth ; the lips have an elevated appearance when the mouth is shut.

Treatment.—Dose of physic, wash out the mouth with astringent solutions, and give sloppy food. If weak and low, give stimulants and tonics, also a little nitre in the water to drink.

FOOT-AND-MOUTH DISEASE.

Symptoms.—The mouth hot and tender, saliva drippin therefrom, partially chewed food is dropped, fetlocks enlarge and painful. Vesicles appear along the mouth, which bur in about twenty-four hours after their formation; and i milk cows, swelling of the udder and tenderness.

Treatment.—Soft sloppy food, opening medicines required, but they must be cautiously given. Wash th mouth with a mild solution of alum, and bathe the feet, an apply mild astringent dressing. Cleanliness must be strictl enforced. Apply disinfectants, such as carbolic acid.

RED WATER.

In Scotland it occurs after calving : but in England an Ireland it takes place at any time, and at any age.

Treatment.—Epsom salts 1 lb., sulphur ¼ lb., croton o ½ drachm, treacle 1 lb.; follow this up with stimulan four times a day, and add ginger or gentian, with wel boiled gruel; give chlorate of potash in water to drink.

RINDERPEST OR CATTLE PLAGUE.

The home of this disease seems to be the southern pa of the Russian empire. It may be defined as an erupti fever of a very contagious nature, due to some poison in tl blood, the exact nature of which is not known.

Sheep present much the same symptoms as cattle.

Treatment.—All remedies have been exhausted for th disease, but with little success; some recommend one thi and some another. It is always a very unwelcome visit and one that requires preventive measures to be strict enforced; for in this course rests our security.

INDEX.

NEILL AND COMPANY, EDINBURGH,
GOVERNMENT BOOK AND LAW PRINTERS FOR SCOTLAND.

William R. Jenkins's
VETERINARY BOOKS.

851 AND 853 SIXTH AVENUE, NEW YORK.

Any of the following books will be sent post paid on receipt of the price ; full Catalogue on application.

PRICE

Animal Castration. By Dr. A. Liautard. 12mo, illustrated $2 00

American Veterinary Review. Edited by Prof. A. Liautard, H.F.R.C.V.S. Issued monthly. Subscription, $3 per year ; single copy............................... 25

Armatage. "Every Man His Own Horse Doctor." In which is embodied Blaine's "Veterinary Art," with 330 original illustrations, colored plates, anatomical drawings, etc. 8vo, half leather..................... 7 50

Armatage's Veterinarian's Pocket Remembrancer. By George Armatage, M.R.C.V.S., with concise directions and memoranda for the treatment in urgent or rare cases. 32mo, cloth................................. 1 25

Armatage. Horse owners' and Stable-men's Guide. Crown 8vo, cloth... 2 00

Baucher. New Method of Horsemanship. Including the Breaking and Training of Horses. 12mo, cloth, illustrated 1 00

Chauveau. The Comparative Anatomy of the Domesticated Animals. By A. Chauveau, Professor at Lyons Veterinary School, France. New edition, translated, enlarged, and revised. By George Fleming, F.R.C.V.S. 8vo, cloth, with 450 illustrations 7 00

PF

PRICE

Fleming. Roaring in Horses. By Dr. George Fleming, F.R.C.V.S. 8vo, cloth, with colored plates........... 2 00

Fleming. On Horseshoeing. By Geo. Fleming. Cloth... 75

Fleming. Operative Veterinary Surgery. By George Fleming. (In Preparation.) Part 1 now ready........ 3 50

Fleming. Propagation of Tuberculosis. By George Fleming. Cloth............................♦............ 2 25

(*)Fleming-Neumann. "PARASITES AND PARASITIC DISEASES OF THE DOMESTICATED ANIMALS." A work which the students of human or veterinary medicine, the sanitarian, agriculturist or breeder or rearer of animals, may refer for full information regarding the external and internal Parasites—vegetable and animal —which attack various species of Domestic Animals. A Treatise by L. G. Neumann, Professor at the National Veterinary School of Toulouse. Translated and edited by George Fleming, C.B., L.L.D., F.R.C.V.S. 873 pages, 365 illustrations, cloth.................... 8 00

Fleming. "Animal Plagues." Their History, Nature, and Prevention. By George Fleming, F.R.C.V.S., etc. Being a Chronological History from the earliest times to 1844. First Series, comprising a History of Animal Plagues from B.C. 1490 to A.D. 1800. 8vo, cloth...... 6 00
Second Series, containing the History from A. D. 1800 to 1844. 8vo, cloth. 4 80

Fleming. Veterinary Obstetrics. Including the Accidents and Diseases incident to Pregnancy, Parturition, and the Early Age in Domesticated Animals. By Geo. Fleming, F.R.C.V.S. With 212 illustrations. 8vo, cl. 6 00

Fleming's Rabies and Hydrophobia. History, Natural Causes, Symptoms, and Prevention. By George Fleming, M.R.C.V.S. 8vo, cloth..... .... 6 00

Hayes. Veterinary Notes for Horse Owners. An Everyday Horse Book. Revised edition, illustrated. By M. H. Hayes. 12mo, cloth......................... 5 00

Heatley. The Horse-owners' Safeguard. A handy Medical Guide for every Horse-owner. By George S. Heatley, V.S. 12mo, cloth................................ 1 50

Hill. The Management and Diseases of the Dog. Containing full instructions for Breeding, Rearing, and Kennelling Dogs. Their different Diseases, embracing Distemper, Mouth, Teeth, Tongue, Gullet, Respiratory Organs, Hepatitis, Indigestions, Gastritis, St. Vitus' Dance, Bowel Diseases, Paralysis, Rheumatism, Fits,

PRICE

Rabies, Skin Diseases, Canker, Diseases of the Limbs, Fractures, Operations, etc. How to detect and how to cure them. Their Medicines, and the Doses in which they can be safely administered. By J. Woodroffe Hill, F.R.C.V.S. 12mo, cloth extra, fully illustrated. .$2 00

Hill. " The Principles and Practice of Bovine Medicine and Surgery." By J. Woodroffe Hill, F.R.C.V.S. Octavo, 664 pages, with 153 Illustrations on wood and 19 full page colored plates. Cloth..................... 10 00
Octavo, 664 pages, sheep............................ 11 50

Holcombe. "Laminitis." A Contribution to Veterinary Pathology. By A. A. Holcombe, V.S. Pamphlet..... 50

Horses and Roads ; or, How to Keep a Horse Sound on his Legs. By "Free Lance"........................... 2 00

Howden. " How to Buy and Sell the Horse." The object of this book is to explain in the simplest manner what constitutes a sound horse from an unsound one. 12mo, cloth....................................... 1 00

Jennings. Horse Training Made Easy. A Practical System of Educating the Horse. By Robert Jennings, V.S. 12mo, cloth.......................... .... 1 25

Jennings. Swine, Sheep, and Poultry. Embracing the History and Varieties of each; Breeding, Management, Disease, etc. By Robert Jennings, V.S. 12mo, cloth. 1 25

Jennings. Cattle and their Diseases; with the best Remedies adapted to their Cure. By Robert Jennings, V.S. 12mo, cloth............................ 1 25

Jennings. "The Horse and his Diseases." By Robert Jennings, V.S. 12mo, cloth........................ 1 25

Journal of Comparative Medicine and Surgery. A Monthly Journal devoted to the Diseases of Animals, particularly of the Horse. Subscriptions, $3 per annum. Single copies, postpaid..................... 30

Laverack. The Setter. By E Laverack. With instructions how to Breed, Rear, Break, etc. Colored illustrations......... .... 2 75

Liautard. Vade Mecum of Equine Anatomy. By A. Liautard, M.D., V.S. 12mo, cloth...................... 2 00

Liautard. "Animal Castration." By Dr. A. Liautard, D.V.S. 12mo, illus. 2 00

Liautard. Translation of Zundel on the Horse's Foot. By Dr. A. Liautard, D.V.S. 8vo, cloth................. 2 00

Law. The Lung Plague of Cattle; Contagious Pleuro-Pneu-

PRICE

monia. Illustrated. By James Law, Professor of Veterinary Medicine in Cornell University. Paper, 100 pp .$0 30

Law. Farmers' Veterinary Adviser. A Guide to the Prevention and Treatment of Disease in Domestic Animals. By James Law, Professor of Veterinary Medicine in Cornell University. Illustrated. 8vo, cloth... 3 00

Lehndorff. Horsebreeding Recollections. By G. Lehndorff. 8vo, cloth... 4 20

Martin. Cattle. Their Various Breeds, Management, and Diseases. By W. C. L. Martin. Revised by W. Raynbird. 16mo, boards... 50

McAlpine. Biological Atlas. Containing 24 plates of 423 colored illustrations. Oblong quarto cloth. By D. McAlpine, F.C.S ... 3 00

McBride. Anatomical Outlines of the Horse. Revised and Enlarged by T. M. Mayer, M.R.C.V.S. With colored illustrations. 12mo, cloth... 3 00

McClure. Diseases of American Horses, Cattle, and Sheep. Their Treatment; with full description of the Medicines employed. By R. McClure, M.D., V.S. 12mo, cl., illus. 2 00

McClure. American Gentlemen's Stable Guide; with the most Approved Methods of Feeding, Grooming, and Managing the Horse. By Robert McClure, M.D., V.S. 12mo, cloth... 1 00

Meyrick. Stable Management and the Prevention of Diseases among Horses in India. By J. J. Meyrick, F.R.C.V.S. 12mo, cloth... 1 00

Miles. Remarks on Horses' Teeth. Addressed to Purchasers. By W. Miles... 60

Moreton. "On Horsebreaking." By Robert Moreton. 12mo, cloth... 50

Moreton's Manual of Pharmacy for the Veterinary Student. By J. W. Morton. 12mo, cloth... 3 50

Navin. "The Explanatory Stock Doctor," for the use of the Farmer, Breeder, and Owner of the Horse. With numerous illustrations. By John Nicholson Navin, V.S. 8vo, sheep... 4 75

Percival. Hyppo-pathology. A Systematic Treatise on the Disorders and Lameness of the Horse. By W. Percival. With many illustrations. 6 vols., boards...34 20

Percival. Lectures on Horses; Their Form and Action. By W. Percival. With eight outline plates. 8vo, cloth. 4 00

Percival's Anatomy of the Horse. By W. Percival. 8vo, cloth... 8 00

PRICE

Reynolds. "Breeding and Management of Draught Horses."
By Richard S. Reynolds, M.R.C.V.S. Crown 8vo, cl.... 1 40
Riley. The Mule. A Treatise on the Breeding, Training,
and Uses to which he may be put. 12mo, cloth, illus.. 1 50
Robertson. The Practice of Equine Medicine. By W.
Robertson... 6 50
(*)Smith. "A MANUAL OF VETERINARY PHYSIOLOGY."
A work distinctive from any other, on the subject
known to the profession, it being exclusively Veterin-
ary and not a Comparative Physiology. By Veterinary
Captain F. Smith, M.R.C.V.S. Author of "A Manual
of Veterinary Hygiene." 8vo, cloth, fully illustrated. 4 25
Steel. A Treatise on the Disease of the Dog. A Manual
of Canine Pathology, Medicine, Surgery, and Thera-
peutics. 8vo, cloth................................... 3 50
Steel. A Treatise on the Diseases of the Ox. Being a
Manual of Bovine Pathology, especially adapted to
Veterinary Practitioners and Students. By John Henry
Steel, M.R.C.V.S., F.Z.S. 8vo. with 118 illus., cl...... 6 00
Steel. "Outlines of Equine Anatomy." A Manual for
the use of Veterinary Students in the Dissecting Room.
By John H. Steel, M.R.C.V.S. 12mo, cloth.......... 3 00
Strangeway. "Veterinary Anatomy." New edition, revised
and edited by I. Vaughn, F.L.S., M.R.C.V.S., with
several hundred illustrations. 8vo, cloth........... 5 00
Stormonth's Manual of Scientific Terms. Especially
referring to those in Botany, Natural History, Medical
and Veterinary Science. By Rev. Jas. Stormonth..... 3 00
Tellor. "Diseases of Live Stock," and their most Efficient
Remedies. By Lloyd V. Tellor. 8vo, cloth, illustrat-
ed, $2.50; sheep...................................... 2 50
Tuson. Pharmacopœia, including Outlines of Materia
Medica and Therapeutics in Veterinary Medicine. By
R. V. Tuson. 12mo, cloth........................... 2 50
Veterinary Diagrams. Five Charts, each 22x28 inches in
size, on stout paper, as follows, sold separately:
No. 1, with eight colored illustrations. External Form
and Elementary Anatomy of the Horse............. 1 50
No. 2. Unsoundness and Defects of the Horse, with
50 woodcuts.. 75
No. 3. The Age of the Domestic Animals, with 42 wood-
cuts.. 75
No. 4. The Shoeing of the Horse, Mule, and Ox, with
59 woodcuts...................................... 75
No. 5. The Elementary Anatomy, Points, and Butcher's
Joints of the Ox, with 17 colored illustrations. With
explanatory text.................................. 1 50
Price per set of five................................. 5 00

PRICE

Walley. "Four Bovine Scourges." (Pleuro-Pneumonia, Foot and Mouth Disease, Cattle Plague, and Tubercle., With an Appendix on the Inspection of Live Animals and Meat. By Thos. Walley, M.R.C.V.S. With 49 colored illus. and numerous woodcuts. 4to, cl........$6 40

Webb. "On the Dog." Its Points, Peculiarities, Instsinct, and Whims. Illustrated with photographs........... 3 00

Williams. Principles and Practice of Veterinary Medicine. New edition, entirely revised. and illustrated with numerous plain and colored plates. By W. Willlams, M.R.C.V.S. 8vo, cloth....................

Williams. Principles and Practice of Veterinary Surgery. New edition, entirely revised, and illustrated with numerous plain and colored plates. By W. Williams, M.R.C.V.S. 8vo, cloth....................

Williams. Chart of the Contagious, Infectious, and Specific Fevers of the Domesticated Animals.................. 1 00

Zundel. "On the Horse's Foot." Translated by A. Liautard, M.D., D.V.S. 12mo, cloth.................... 2 00

VETERINARY BOOKS IN FRENCH.

Benion. Traité de l'Élevage et des Maladies des Animaux et des Oiseaux de Basse-Cour........................$2 80

Benion. Traité de l'Élevage et des Maladies du Mouton.. 3 60

Benion. Traité de l'Elevage et des Maladies du Porc..... 2 60

Beugnot. Dictionnaire usuel de Chirurgie et de Médecine Vétérinaire. 2 forts volumes in-8, avec planches....... 7 20

Bouley. La Rage, moyen d'en éviter les Dangers et de prévenir sa Propagation............................ 40

Bouley-Reynal. Nouveau Dictionnaire Pratique de Médecine, de Chirurgie et Hygiène Vétérinaire (to be completed in 18 volumes), chaque volume............... 3 00

Colin. Traité de Physiologie Comparée des Animaux; Par G. Colin, Professeur à l'école Vétérinaire d'Alfort; avec Figures intercalées dans le texte. 2 vols. in-8....10 40

Cruzel. Des Maladies de l'Espèce Bovine. Par J. Cruzel. 5 60

Dictionnaire. Lexicographique et Descriptif des Sciences Médicales et Vétérinaires. Un très-fort vol. de plus de 1500 pages.. 8 00

Gourdon. Traité de la Castration des Animaux Domestiques 3 60